OUR BIRD FRIENDS AND FOES

1. Ruby-throated Humming Bird. 2. Migrant Shrike. 3. Bobolink. 4. Robin. 5. Mourning Dove. 6. Cliff Swallow. 7. Nighthawk. 8. English Sparrow. 9. House Wren. 10. Redstart. 11. Crow. 12. Great Blue Heron. 13. Barn Owl. 14. Mocking Bird. 15. Bald Eagle. 16. Herring Gull. 17. Black-capped Chickadee. 18. Jungle Fowl. 19. Bobwhite. 20. Red-headed Woodpecker. 21. Mallard Duck. 22. Emperor Penguin. 23. Golden Plover. 24. African Ostrich.

Our Bird Friends and Foes

By

WILLIAM ATHERTON DuPUY

Foreword by

WALDO LEE McATEE

Introduction by

ROBERT RIDGWAY

Illustrated by

GEORGE MIKSCH SUTTON

DOVER PUBLICATIONS, INC., NEW YORK

Published in Canada by General Publishing Company, Ltd., 30 Lesmill Road, Don Mills, Toronto, Ontario.
Published in the United Kingdom by Constable and Company, Ltd., 10 Orange Street, London WC 2.

This Dover edition, first published in 1969, is an unabridged and unaltered republication of the revised (1940) edition of the work originally published by The John C. Winston Company in 1925, except that in the present edition the frontispiece is reproduced in black and white, whereas formerly it was in color.

Standard Book Number: 486-22269-1
Library of Congress Catalog Card Number 69-17525

Manufactured in the United States of America
Dover Publications, Inc.
180 Varick Street
New York, N. Y. 10014

PREFACE

MOST people will agree that birds constitute the branch of the animal kingdom possessing greater beauty and charm than any other. One need but point to the exquisite use of color made by birds, to their undisputed mastery in music, to make out a case.

Yet, so shy are birds that few people ever come to know them.

My purpose here is to take the reader into birdland, to acquaint him with much that goes on there, to show him the routine of bird life, its hardships, its joys, its romance, all sketched into just enough of scientific background, relieved of its technical terms, to give him a basis of real understanding.

No one man can learn enough of birds at first hand in a single lifetime to do justice to the subject. It is only as "heirs to all the ages" that the facts are massed for us. I have, by dint of much reading, borrowed facts from many sources. I have, in addition, been so fortunate as to come in contact with a few men who, in their generation, are outstanding in their knowledge of birds, and they have been generously helpful.

It is probably true that no city in the world harbors so large a colony of scientific men as does Washington, and it is there that I have worked. No government employs so many specialists in the ramification of government bureaus as does the United States. No group of men gives itself more unstintedly to the cause of education.

One such man who has helped me to know birds is Dr. Charles W. Richmond, for thirty-five years an ornithologist with the United States National Museum. He has left the home field to other men and has studied those birds whose habitats are beyond the bounds of the United States. To him the bird map of the world unrolls as a simple scroll. He has counseled with me, has read and edited my manuscript chapter by chapter as it has been written.

I have relied much, also, on the judgment of another scientist of the government service who has given his life to the study of birds, approaching it, however, from quite another standpoint. Waldo Lee McAtee studies the food habits of birds for the Biological Survey, which is in the Department of Agriculture. That department wants to know which birds are helpful and which are harmful from the farmer's standpoint. The determination of this fact depends on finding out what they eat. Mr. McAtee is the government's star student of the food habits of birds. He has studied the subject longer than any man in the world, and undoubtedly knows more about it than any other. But in studying food habits he has associated long and intimately with birds and acquired charming intimacies with them.

Mr. McAtee has been another of my consultants. He has given my manuscript a second reading in the interest of accuracy. He has also been so good as to write a bit of a foreword to the book.

The dean of American ornithologists, however, is Robert Ridgway, likewise in the service of the government, the man who has the rare distinction of having guided the National Museum in its bird studies for the past fifty-one years. He is Curator of the Division of Birds of the National Museum. Beginning as far back as 1867, he has devoted himself unremittingly to the task of piling up always more information on the subject of the birds of North America. As a young man he wrote, in collaboration, five great volumes called *History of North American Birds.* His later eight huge tomes, written under the title *Birds of North and Middle America,* are monumental works. He has done much of that thing dear to the heart of the scientist, the development of facts not before known. Thus he has added much to the sum total of human knowledge. He named more than five hundred birds that had not before been known to the scientific world. No other man ever contributed as much to the knowledge of American birds as Robert Ridgway. No other man ever can, because the opportunity no longer exists.

It has been my final great pleasure that Mr. Ridgway should read and correct my manuscript, that it should meet his approval, and that he should honor me by writing an introduction to my book.

The modest writer who has wrestled with the problem of making the bird world easier of access for the casual reader gains reassurance when such men as these lend a hand to his task. That the premier governmental authority on the food habits of birds should write a foreword, that the man who has guided the National Museum in its bird studies for half a century should write an introduction to his book, is sufficient reward to the author for his effort. But, in addition, it gives him a confidence, in presenting the result to the reader, that he has been as carefully guarded against mistakes and misrepresentations as it is humanly possible that he should be.

WILLIAM ATHERTON DU PUY

FOREWORD

THE bird branch of the animal tree, since forking off from its reptilian stem, has grown much and flowered beautifully. How much more admirable, for instance, is the soft plumage of the bird than the harsh body mail of the reptile. How much more pleasing by contrast are the notes of birds, some of them among the best music in the world, to the glum silence of reptiles, broken only by threatening hissing.

Birds have a monopoly of feathers; every feathered animal is a bird. No bird lacks feathers; no other creature has them. Feathers not only have proved wonderfully adaptable in aiding the bird to fly, but they have developed as down and thickly set body plumage, that enable the bird to resist cold, to float like a cork on icy waters, and even to repose on the ice itself, enveloped in a thick downy mantle that shuts out the killing frost.

Birds made a great and wonderful advance when they left behind that reptilian inheritance of cold blood, which bound them to the tropics, and began to develop a warm-blooded circulation. Now their systems are so efficient in keeping up body temperature that birds are the warmest of all living things.

To maintain their great activity and high temperature birds must have nourishing food and plenty of it. Young nestling birds, as a rule, consume as much as, or more than, their own weight of food daily, and the adults are gross feeders. The phrase "a bird-like appetite," meaning a small or delicate one, comes from a misunderstanding of the facts.

The insects are now acclaimed as man's most serious competitors for the possession of the earth. There are more of them than of all other animals put together. They are found everywhere, and are in convenient sized packages to serve as morsels for bird dinners. Birds prey upon them. In fact, birds have been widely and diversely specialized to search out and destroy insects of practically every kind. Crows, blackbirds, larks, thrushes, and numerous other birds devour insects living on or near the surface of the ground; flycatchers, swifts, and swallows catch them in the air; vireos and warblers search the leaves and buds of trees; chickadees and kinglets scan the buds and twigs; creepers and nuthatches patrol the bark; and woodpeckers dig into the trees and extract woodborers even from that fancied security. The water-inhabiting insects are preyed upon by still other birds, and the night-fliers have no immunity, for nighthawks, whippoorwills, and owls begin their warfare on the insect hosts as birds of the day take their repose.

Scarcely an insect pest exists that does not have its bird enemies, and so effective are the attacks of birds that insects are sometimes noticeably reduced over large areas, and entirely destroyed locally. The farmer, the fruit grower, and the forester alike have the birds to thank for valiant aid in combating their insect foes. The potato beetle has been entirely eliminated from some patches by bobwhites. Tent caterpillars have been partially exterminated in some areas by cedar waxwings.

Woodpeckers have been observed to destroy bark beetles to the extent of half or more of the brood in groups of trees. In Nova Scotia a single kind of bird—the red-eyed vireo—has been known to devour in various years from ten to ninety per cent of the larvæ of the fall web-worm.

Among other forest pests the notorious gipsy moth is preyed upon by forty-six kinds of birds, and its destructive relative, the brown-tailed moth, by thirty-one. Insect foes of the orchard that many birds like are the codling moth, eaten by thirty-six species of birds which sometimes destroy from sixty-six to eighty-five per cent of the wintering larvæ of this pest. Scale insects which spell perdition to trees are known to be fed upon by some seventy different kinds of birds, and cankerworms, the especial blight of the orchard, by seventy-six. Well-known farm pests and the number of their known bird enemies are: cotton worm, forty-one; cotton boll weevil, sixty-six; army worm, forty-three; alfalfa weevil, forty-five; white grubs, eighty-one; cutworms, ninety-eight; and wire-worms, one hundred seventy.

Birds appear to be the most efficient force that nature has developed to check undue multiplication of our insect foes; they have demonstrated over and over again their great usefulness, and they should be so protected and encouraged that their services will reach the maximum value to mankind.

I have carefully read the chapters of this book and have lent a hand here and there toward insuring its technical accuracy. I feel that the author has succeeded to an unusual degree in crystallizing the interest and romance of bird life, in interweaving with sprightly accounts of pivotal species of birds informative discussion of the leading trends in bird evolution. Books that vitalize such subjects, that make them easy reading, and at the same time maintain a background of the proper scientific structure, are very rare indeed. This work will help to consolidate the advances already made in winning public sentiment to a real appreciation of and to an intelligent conservation of bird life.

WALDO LEE MCATEE
Assistant in Charge, Division of Food Habits
Research, U. S. Biological Survey

CONTENTS

ILLUSTRATIONS

INTRODUCTION

MAN, self-centered and egotistical, is prone to overlook the fact that other created things in the world besides himself deserve consideration. He rarely realizes that he is, in truth, only one of countless living forms, all placed here by the same creative force for some useful purpose, and with all of which he is concerned in many ways affecting his welfare. Both the plant world and the animal world afford him food and raiment, as well as practically every comfort and luxury. In both are his worst enemies and his best friends; and it should be a part of his business to learn carefully to distinguish between those that he should destroy and those that he should protect. This he can learn only by careful observation or through the observation of others. The study of Natural History is, therefore, not only rational, but eminently practical.

The view that, of all forms of life on the earth, man is paramount is, of course, entirely justified; but the contention that all else was placed here solely for him to do with as he likes is a fallacy. The riches of Nature were given man for his use, not for abuse; to take toll of for his necessities and comfort, not to be wasted by ruthless destruction, or exploited for the amassing of wealth. His own interests demand that he should use the rich resources with which he has been blessed with intelligence and moderation; but, unfortunately, he does not, and man has, by his reckless destructiveness and wastefulness, earned for himself the unenviable distinction of being, of all creatures, the one Great Destroyer.

So complex are the interrelations between the countless forms of life that they react, one way or another, on each other. Probably, however, no single creature is essential to the existence of the remainder; certainly even man himself, should he cease to exist, would be missed by the animal world only as would a scourge which had long devastated a human community be missed when it eventually disappeared.

The more one studies Nature, the more one must be impressed by the lessons thus learned; with the wonder of it all, and with the feeling that only a divine origin and purpose can account for the marvels, the mysteries, the miracles that the study reveals. The transformation from the egg to the perfect bird; from the ugly caterpillar to the lovely butterfly; from the seed, seemingly as lifeless as a pebble, to the fully developed and noble tree—these are but a few indeed of the wonders of Nature.

In spite of some who in their conceit imagine that man has a complete monopoly of reason, it can be, indeed has been, demonstrated that animals lower in the scale of development do, to a greater or less extent, possess both intelligence and reason.

There is, for example, the unerring accuracy of bees and wasps in returning to their homes and the entirely humanlike method by which they get their bearings. There is the perfect organization of ant colonies, each kind in its own special way, their division into workers, soldiers, nurses, etc.—organizations and methods as perfect and orderly as those of the most advanced of human communities. Some birds show what appears very like reason when they build their nests in bushes or trees inhabited by colonies of the dreaded fire-ant, and others when they build them close to the nests of pugnacious and formidable wasps, where they themselves are not molested, but where no marauding opossum, monkey, or even human being dares to venture. Bower-birds, of Australia, build playhouses and collect brightly colored objects, much as certain human beings collect curios. The garden-bird of New Guinea constructs a playhouse with an imitation garden, covering the ground about its hut with a carpet of green moss which it decorates with brightly colored flowers, as regularly as a housewife might freshen the bouquets in her vases.

In countries that are sparsely inhabited by man or that have not long been occupied by him, there has become established an equilibrium called the balance of Nature. This results from the check put on one species by another, so that none can increase at the expense of others. For example, here in the United States rabbits were kept in check by foxes, wild cats, and other carnivorous animals; destructive rodents were kept within bounds by the same animals, by skunks, and, especially, by hawks and owls. But wherever civilized man appears, he proceeds to kill off the foxes, wild cats, skunks, hawks, and owls—many other beneficial creatures besides—rabbits and rodents increase in proportion to the decrease of their natural enemies, and thus the balance of nature is upset or destroyed.

Man, the Destroyer, is, indeed, responsible for serious

sins of this kind. Partly from ignorance, but mostly on account of the survival, from his early barbarism, of the killing instinct, no bird, conspicuous on account of size or appearance, is safe from the man creature carrying a gun. Thus some of our most beautiful and interesting birds have been completely wiped out of existence. One such was the passenger pigeon which, within the memory of persons now living, was undoubtedly the most numerous bird in the world. Other birds are rapidly disappearing, indeed have quite disappeared from much the greater part of the area which they once inhabited, as, for example, that noblest of all game birds, the wild turkey. Unless effective means are taken to check this slaughter, the next generation will certainly miss some of the species that still, though in greatly reduced numbers, exist.

In the wholesale destruction of wild life which has taken place, only the insects have escaped; and of all enemies with which man has to contend none can compare with the members of the insect world, at once the most numerous and voracious of all creatures. Now among the enemies of insects the birds are first, because the food of most birds consists chiefly or to a greater or less extent of insects. The digestive processes of birds are extremely rapid, and so it comes to pass that every insect-eating bird each day devours more than its own weight of these arch-enemies of the farmer and horticulturist. It is safe to say that but for the help thus rendered by the birds it would hardly be possible to grow crops of grain, forage, or fruits. The sooner more people realize this fact the better it will be for all.

It is now too late to undo the harm that has been done

through thoughtlessness, ignorance, and the mistaken belief that our resources were inexhaustible, but there is yet time to make some amends by earnest efforts to awaken the people to a full realization of the dangers that confront them. The study of birds has grown a good deal in the present generation. In my younger days, some sixty-five years ago, the number of persons in the United States who were known as ornithologists could be counted on the fingers, alone, of one hand. Twenty years later, when the American Ornithologists' Union was organized, 134 names were included in the membership. Three years later the membership numbered 158; while in 1924 it numbered 1395. These numbers represent but a small proportion of those who are seriously interested in the study of birds, for there are other ornithological societies, some of them including a very large number of members, and there are thousands who are not yet affiliated with any organization.

It is, however, no easy task to get an understanding of birds to the masses. The subject has been ably discussed, but these discussions have reached but the few. The great problem is that of reaching the multitude, and this, it seems to me, may be accomplished through the schools. The interest in Natural History is, obviously, growing, and there is reason for hope that it may gain a firmer grip on the attention of the rising generation.

The greatest contribution to an understanding of birds probably lies in the creation of those books that carry the message most readily to the popular mind.

Many books on birds, intended for those who are not specialists in the study of ornithology, have been published. Some of them are entertaining, but full of error; others are neither interesting nor accurate, while some are too technical or "dry" to be readable. This book of Mr. Du Puy's, of which I have read every word of the manuscript, has none of these faults. On the other hand, it possesses an excellence peculiarly its own—a charm of style, a wealth of accurate information, that no other book on birds that I have read can equal. I believe that if it could be put into the hands of everyone who cares to read on this subject, if it could be got into every school in our country, one of the most serious, if not the most serious, of the problems which confront us as a nation would be solved.

ROBERT RIDGWAY
Curator of Birds, U. S. National Museum

Chapter I

THE GOLDEN PLOVER

A GREAT transport, bearing troops, was five hundred miles out of Honolulu bound for San Francisco. There in that mighty waste of water, which stretches for distances as great as from New York to New Orleans without a single island to break its surface, this transport met a frail voyager, a golden plover, laboring heavily on weary wings. Well might it be weary, for to reach a spot so remote from land had required days and nights of steady flying without one moment of rest. The ocean beneath was a deadly menace, for this was not a swimming bird, and could never rise again should it once alight.

The timid, tired, plover turned from its course and followed the transport for two days and two nights, keeping above or near it, but unable to muster courage enough to alight upon its deck to rest. In doing so it traveled much longer and farther than if it had gone on to Hawaii. Occasionally it would fly high into the air, almost out of sight, evidently that it might look into the distances on the chance of sighting land. Then it would return and flutter again about the ship, uttering its plaintive cry, but never brave enough to fly to that deck which offered rest. Finally its weary wings gave way, it dropped into the water, and the mighty ship plowed on.

This chance meeting at sea gives a glimpse of one phase of one of the most remarkable life stories of the birds that inhabit the globe. This golden plover, far at sea, was but a straggler from one of the flocks, thousands strong, that every year make that long flight, without stopping, from Alaska in the far North to the Hawaiian Islands in the mid-Pacific. These islands, which are its winter home, are the most isolated bits of land on the face of the earth. From them it is two thousand miles to San Francisco to the east, twenty-four hundred miles to Alaska to the north, and thirty-five hundred miles to Japan to the west. They stand alone near the center of the greatest waste of water in all the world.

On the Hawaiian Islands there live for two-thirds of each year these flocks of golden plover, members of a race which, in some of its branches, provides shy visitors, upon occasion, to lowland meadows all around the world. In Hawaii, unexpectedly, of an April evening, some observer may see flocks of these plover take to the air and circle about. Finally, they seem to get their bearings, and strike out unerringly for the North.

They have begun a journey in which there can be no break until they have covered that immense stretch of twenty-four hundred miles to Alaska. Swift of wing are the plover, and it is estimated that on this journey they travel at a speed of forty miles an hour. At this rate, under favorable conditions and without meeting any adverse winds, there are sixty uninterrupted, foodless, drinkless hours of flying required to reach the outlying islands of Alaska.

Even then the journey of the plover is by no means

ended. After a period of rest the birds push on until they reach that fringe of land skirting the Bering Sea and the Arctic Ocean which wakes for a brief period in each mid-summer, shakes off its blanket of snow and, through the long Arctic day, develops a wealth of flowering plants and shelters myriads of birds such as plover and their allies. These important bird homes are the tundras of the far North and here the golden plover lays its eggs and rears its young.

The golden plover is a wader, a shore bird, a relative of the snipe, the lapwing, the curlew, the woodcock, and more distantly, of the crane and the stork. A familiar near relative is the killdeer, a plover found almost everywhere in the United States. The killdeer, however, migrates by comparatively short stages, overland, taking no such ocean flights as the golden plover.

No golden plover ever sat on a limb, a fact that has written itself strangely on the body of this bird. The plover has but three toes, all of which point to the front. Most of the other birds have four, one, like a thumb, pointing backward and making it possible for it to grasp a limb. The plover, therefore, keeps to the ground more than some of its relatives which still retain the fourth toe. It is chiefly a marsh bird living in damp places and given to wading. However, it is not so developed as a marsh inhabitant as the snipe, but has a short bill with which it picks up insects and worms on the surface rather than a long one with which to probe the mud as does its relative.

It is a handsome bird, black above, spotted with golden yellow, black underneath in summer. with white flanks.

It is nearly eleven inches long and its wings spread almost two feet from tip to tip. Its body is about the size of that of a mourning dove, plump, because of the great

IMMATURE GOLDEN PLOVER

development of its flying muscles, and very good to eat. In Europe it is known as the rain bird, being noisy ahead of bad weather. The name plover comes from "pluvia," which means "rain."

In summer, during the long day of the northern tundra, the mother golden plover makes her moss-lined nest on the ground and in it lays four buffy eggs, heavily marked with chocolate spots, while her mate circles attentively overhead, uttering constantly his "taludle, taludle, taludle" cry. Later, however, he is mustered into more practical service and made to sit on the nest, for Mrs. Plover is a modern woman.

The tundras are strange waste lands of the North, vast marshes on which the snow melts faster than it can be absorbed by the ground.

Here is the greatest abundance of mosquitoes and gnats that occurs anywhere, such swarms of them as to make it a difficult place of residence for man or beast. But gnats and mosquitoes do not disturb the feathered plover. On the contrary, they provide endless quantities of food and further help the birds by insuring freedom from beasts of prey.

It is fortunate that in a few brief weeks the new generation of plover grows lusty and strong, for the long Arctic night soon begins to shut down, bringing with it cold and darkness which transform this end of the world into a forbidding and almost foodless and lightless waste. Before this occurs the plover again take wing. Lusty and fat they drift to the South, this branch of the family finally reaching the shores of the Pacific. There they are again faced with that long flight of sixty foodless, drinkless hours back to the Hawaiian Islands. Again they leap courageously into the air, circle about, get their bearing, and strike out unerringly, after four months' absence, for their winter home.

It is a marvelous fact that, by instinct, these birds can cross the trackless ocean and find their home in the midst of it as unfailingly as can a ship equipped with all the instruments which man in his wisdom has been able to devise. In September the Hawaiian Islands are aswarm again with the golden plover that flew away in April or May, and with their numerous young that hatched from speckled eggs in the northland, all busy serving man by eating armyworms and cutworms that injure the sugar cane crop.

The plover are paler, yellower now than when they left in the spring, having changed the black feathers of their undersides for white ones. This change of dress at different seasons is a habit of many birds and one that often puzzles man observers. The plover is golden yellow in the autumn like the leaves, the grain, and the goldenrod.

Hawaii is not the only land that furnishes its quota of feathered summer tourists that fly away to the Arctic. All around the world during the month of May the plover travel routes are crowded with birds flying to the North. From the islands of the Indian Ocean, Australia, New Zealand, from Persia, from the waste lands of Africa, from the heart of Brazil, caravans put forth to the North. Uncounted thousands of plover form a belt around the North Pole for that brief season when the midnight sun bathes the region in warmth. Uncounted thousands of young plover are reared there and fly away to their winter homes, visiting nearly every land in all the world.

One of the most famous groups of plover, a little different from those of the Pacific, is that which every year passes through the United States. It has its winter home in the waste lands of Brazil, and in the pampas of Argentina, where the land is low and swampy. Here the flocks find abundant food and are, we hope, not too much disturbed by the Gauchos and their herds.

Then a time comes when the thought of nesting enters their minds and the flocks start northward. They follow the continent to Panama, up through Central America, to the point where Yucatan sticks out into the Gulf of Mexico. They cover the Gulf, five hundred miles, at a flight, and travel northward through the Great Plains areas of the United States. On they go to the northward, spreading both to the east and the west, until they reach the Arctic, where they breed along the whole northern shore of the continent. Here they feed on that limitless supply of gnats and mosquitoes of the far North, other insects, and the quick-ripening berries.

When the young have grown strong they take the southward route by the land east of Hudson Bay and are likely to strike the Atlantic Ocean at Nova Scotia. This

A GOLDEN PLOVER MOTHER AND HER LITTLE ONES

group of plover may make Nova Scotia the starting point for an air flight straight to the South, twenty-five hundred miles without a stop, the longest continuous flight ever made by land birds; a flight that may not be interrupted until the mainland of South America is reached. Then these plover go on across the greater part of another continent and find their old winter haunts in Brazil and Argentina.

It is the habit of most birds to migrate. They have found that the North is roomier, richer in food, and generally a pleasanter place in which to dwell in the summer time, and that the South is more delightful when winter comes. So easy is it for the birds to travel that they long ago adopted a program of following the seasons back and forth from north to south, from south to north.

They did not all learn at one time of the utility of migration, so it happens that different birds move by various routes and at different times. The robin, for instance, can live in winter in a climate which ranges around forty degrees in temperature. So it does not go very far south and follows the melting snow to the northward and becomes one of the first harbingers of spring.

The cliff swallow and the black-poll warbler dwell as close neighbors in Venezuela in the winter time. When spring comes on, however, it is odd to note the different courses that they may follow. The warbler may strike out, flying at night, to the northward by way of Cuba and Florida. Then it turns to the left, goes on and on to the northwest, until, by the end of May, it is building its nest in the solemn pines where the Yukon River flows into Bering Sea.

The cliff swallows start northward at the same time, but fly always by day. Instead of striking across the Caribbean Sea they follow the land around the curve of the Gulf of Mexico, where some of them bear to the right, cross the warbler current, and strike up the Appalachian chain through the United States, and on to build their nests in the cliffs where the St. Lawrence empties into the North Atlantic. So may these two winter neighbors take different courses to the north and put the width of a continent between them when they come to nest building.

Few birds travel distances as great as those flown by the golden plover. Of all the land birds it is probably the greatest traveler. A bird which can rest upon the ocean, however, has an advantage in long distance travel over one that cannot, and so it comes about that the champion long-distance journeyer of them all is a water bird, the Arctic tern. This little cousin to the gulls nests even beyond the tundras which are the summer home of the golden plover. It has been found within eight degrees of the North Pole, hatching its young almost in the snow. It lives here in the far North as long as the sun never sinks below the horizon, but as the Arctic night approaches strikes out to the south. It travels no less than eleven thousand miles and arrives in the Antarctic by the time the long day is there setting in. This bird, which lives most of its life at the ends of the earth, gets more sunshine than does any other creature. It is known as the bird that shuns darkness and it does so by traveling twenty-two thousand miles a year.

To the great mass of birds living on the western hemisphere the arrangement of its land areas is quite satisfactory. North America widens out in its cooler latitudes and offers much pleasant summer nesting space. The mass of the bird population summers in North America. When the drift to the south comes in the autumn, the southern United States, Mexico, Central America, the West Indies, offer winter homes for many of them, but vast numbers of them also drive on to the tropics or even farther in America. That belt on the map which covers the narrow strip of land that connects these two continents, and which includes the groups of islands to the east, is the great travel belt of the birds. Those that summer in the western part of the continent are likely to follow the land route to South America. Most of the birds, however, summer east of the Rockies, and, going south, strike the Gulf of Mexico. Across that Gulf to Yucatan is a flight of from five to seven hundred miles. The great majority of the migrating birds go straight across the Gulf. Many follow the peninsula of Florida, hop across to Cuba, to Jamaica, and on down the island chain to South America. Only the golden plover is bold enough to desert the land route and take the short cut from Nova Scotia to the South American mainland, twenty-five hundred miles by an air line.

One of the oddest things about the long journeys taken by the golden plover is the fact that these journeys in many cases seem entirely unnecessary. In Hawaii, for instance, when the plover starts on its long journey to Alaska, there seems to be no reason for it. Those mid-Pacific islands are pleasant places in which to dwell in the summer time, the supply of insect food is abundant, and the plover might just as well make their nests here and save themselves the hardship of those great flights back and forth to the polar regions.

Students of bird life have tried to find out what it is that urges this shy, shore bird to go on so long a pilgrimage. Most of them believe that the plover originated in the far North at a time when that region was much warmer than it is now. As the North grew colder the bird was driven south in winter for food. The Arctic, however, was its real home, its nesting place. Conditions there in the summer time remained favorable and it always went back to rear its young. It continued to do this as centuries passed and its journeys lengthened to the south. For hundreds of thousands of years these birds have flown back and forth, half across the world. They were doing it possibly before the man-race came into being, before some of the continents had taken on their present form.

In the heart of Russia, for instance, there are vast lowlands. The plover do not fly straight across these lowlands, but make a huge half-moon sweep around them. This sweep follows what the geologists say was once a shore line, when the plains were a part of the ocean. It was then the birds got the habit of going this way.

Probably the most remarkable example of the instinct to go south in winter was given by the migratory quail which was introduced into New England from Europe. When autumn came the young quails, born in America, took wing for the south, flew straight down into the Atlantic Ocean until they were exhausted, dropped into the water, and were drowned.

It is a strange thing that the young of some shore birds,

which, of course, have never made a migration flight, do not leave for the south until after the old ones and that they seem to follow these routes without difficulty. This is explained by saying that it is instinct with them.

It used to be that plover were very abundant in the United States during their travel season. There was a time when a hundred plover could be bought for half a dollar in the markets of Chicago. This was years ago when great flocks of these birds were being killed by market hunters. They shot the birds as they alighted in multitudes on the meadow and farm lands, stopping on their journey for food.

This food consists almost entirely of insects, as the delight of the plover is grasshoppers and caterpillars. It has happened in the prairie states that vast swarms of plover have come to the aid of the farmers in time of grasshopper plagues and saved their crops by devouring these pests. The birds feed also on a number of other insect enemies of agriculture, including the chinch bug, potato beetle, squash bug, clover-leaf and clover-root weevils, bill-bugs, and the cornfield ant.

Plovers render only service to the farmer, since they eat nothing that he would have them spare, but live entirely upon those insects which would devour his crops. Thus it has always been true that when flocks of plover, or prairie pigeons, as he is likely to call them, settled upon his fields, they were sure to be doing the farmer a great deal of good. They were eating their fill of insects which are the worst enemies of his crops. The farmer, had he been wise, would have offered every encouragement to the plover to come again. Instead of this, however, he has usually gone forth with his shot gun and slaughtered great numbers of these unsuspecting children of the North.

A flock of golden plover might see the farmer coming, might sit still, and wait his close approach. Then, as though on signal, they would rise in a dense mass and fly away in perfect formation. The gunner might shoot through them and bring down scores of birds, yet the flock, little suspecting the slaughter that threatened, might circle about and again fly over him, again paying the same deadly toll. Thus were plover easy victims to the mass shooting of the farmer.

A scheme for netting plover also led to their wholesale slaughter. Out upon some desolate flat the trapper would arrange a most ingenious device. He would find a patch of shallow water, and in this would build his trap. Across the middle of it he would throw up a long narrow island and over this island would suspend his net. In the shallow water near it he would build a number of tiny islets, not bigger than your hat, and on each of these would tie a captive plover.

When the migrating flocks came along they would circle shyly over such a flat, wondering if it would be a safe place to seek food. Always, before alighting, they would fly a number of times back and forth. If, as they did so, they should hear the piping call of one of their kind on the ground below, they would take this as proof that here was a safe place to alight. Had not the other plover already settled here?

They would swoop down to the spot from which the voices of the tethered plover came and seek to alight about them. The land that offered the nearest sitting place for them was that on the long narrow island in the shallow water. Some of them would settle here and find, to their joy, that this island was well provided with worms. The flock would fly back and forth and more and more of them would light in this strip of land. Finally, when it was well covered, the plover trapper, hidden near by, would pull a rope which would release the suspended net and it would fall upon the little island and envelop all the plover that had alighted. Thus were they captured wholesale.

By these and other devices the slaughter of plover was great and the birds were greatly reduced in numbers. In addition to these losses, casualities of their long migration trips and shooting again in South America depleted their ranks, until in recent years the birds have become very scarce and are even threatened with extinction. Since they are in every way beneficial to man, and in no way harmful, it is being urged by those who study such situations that the plover should be entirely protected from the huntsman. If this were done they would slowly increase in numbers and come to play a very useful part in suppressing the increasing insect plagues which threaten the welfare of the world.

QUESTIONS

1. How did it happen that the tired plover was met by the steamer far out in the Pacific? How many hours do these birds fly without stopping?
2. Describe the spring flight of the plovers from Hawaii. Where do they go? What for?
3. To what class of birds does the plover belong? Which of its relatives do you know? In what sort of country do they live?

4. Describe the plover's color scheme. What odd thing is there about its feet?
5. Why are the tundras good places for plover nests? What do you think must happen among the plover in this far North when the long night begins to shut down?
6. How, do you figure, are these birds able to find their way back to these mid-Pacific islands?
7. Where else in the world, besides Hawaii, are plover found in the winter? Describe the flight north of the plover of the world and the ring that they form around the pole.
8. Where do the plover live that cross the United States in their migration? Trace on the map the journey they make every year.
9. Many birds go north at nesting time. What are the practical reasons for this? Tell of the surprising journeys some of them make.
10. Tell the story of the bird that shuns darkness.
11. What is the main line of travel back and forth from the tropics to the north? Trace the Florida route.
12. It appears that many of the birds could stay in the South and raise their young quite comfortably. What is there in their race history that causes them to come north for the nesting?
13. How do plover render a service to the farmer? Describe some of the methods of killing them. What should be the attitude of the citizen to these birds? What would be the result?

CHAPTER II

THE SEA GULL

AN observer sitting on a rocky beach in Labrador, or Patagonia, or Sicily, might witness the following trick, indicating bird intelligence. A sea gull will fly up from the water's edge carrying some object in its beak. When the gull is twenty or thirty feet above the ground it will drop this object upon the rocks. It will carefully follow it down and examine it. It may take the object again into the air, this time to an elevation of forty or fifty feet, and again drop it. Once more the gull may return to the beach to inspect the object, again fly into the air, this time going up maybe seventy feet, and again drop the thing it carries.

Coming back to the ground this time it screams joyfully and is soon engaged in its favorite pastime of eating, for the gull is one of the greediest of birds. An examination will show, there among the rocks, the cracked shell of a mussel with the contents gone. This gull had carried the mussel to one height, then another, each time dropping it, until finally the fall had been great enough to burst it open and had given the wise bird its breakfast.

Again, an observer on one of the mud islands off the Louisiana coast, may have watched a huge pelican splash into the water and come up with a fish in the big pouch beneath its beak. No sooner was this capture made than a laughing sea gull began hovering over the pelican, sometimes even alighting on its beak or head. The stupid pelican seemed not to mind. The gull, however, knew what it was about. It knew that, if the head of the captured fish was pointed down the pelican's throat, it

LAUGHING GULL STEALING A FISH FROM A BROWN PELICAN

would be promptly swallowed. If, on the other hand, its tail was pointed down, the bird could not swallow it. In such a case the pelican would have to turn the fish around. To do so it would have to throw it out of the pouch, flip it over, and start it down the right way. This the pelican can do very cleverly.

The time when the fish was in the air was the moment of

opportunity for the gull. It darted in, seized the prize, and was away, screaming joyously.

A visitor to the water front may have thrown to one of these gulls, which will eat almost anything, a piece of salt fish. The gull would have grabbed it, and half swallowed it, before realizing that it was covered with this salt, which is very distasteful to the bird. Having made this discovery the gull would spit the fish out in disgust. The morsel still appeared attractive, however, so the gull would examine it carefully. Finally it would take it to the water's edge, and wash it thoroughly, then start again to swallow it, but it would still be too salty. It would then take it back to the water, wash it for ten minutes, then find it acceptable, and finally bolt it.

These are evidences of wisdom on the part of the sea gull, but despite its knowingness in certain directions, it is a very poor mathematician. It cannot count. One may take advantage of its weakness in figures to find out the secrets of its private life.

Visit some nesting place of the gulls, some rocky islet off the coast or in some of the larger inland lakes, taking with you a small tent and two companions. Pitch the tent among the gulls, and, with your two companions, go inside it. Where three had entered, two may come out and walk away. The gulls, being poor at counting, will believe that all have gone and that they are alone. They will take their places close about this tent and go about their business as though it were not there.

You may cut peep holes in the sides of the tent big enough to show the whole of your face without arousing suspicion. If a gull, not six feet away, looks up and sees that face it pays no attention. It does not recognize this mere part of a human. A face to it, without the accompanying body, arms, and legs, is not a man.

Here in the sea gull colony the nests are but two or three feet apart, but each is a separate establishment. Each male gull and his wife is raising a family of two or three chicks, guarding it jealously, feeding it abundantly. Here the mother is watchful of her big-headed, awkward, down-covered, goblin-like children. If one of them attempts to steal away she pursues it and whips it soundly. If a tiny gull from some neighboring home appears she beats it most cruelly, often actually killing it. In the colonies of certain kinds of gulls the number of young ones that are thus killed by old gulls is surprisingly high. The young of the sea gulls must behave very properly or suffer such consequences.

The grown-up gulls are very quarrelsome among themselves and if one such approaches the nest of another, a desperate fight is likely to take place. Two of these female gulls will lock bills and sway back and forth, at the same time beating each other soundly with their wings.

There is a word in the English language, "gullible," meaning easily fooled, and derived from the name of this bird. Its presence in the language indicates that the gull was once thought to be simple-minded. The early English were evidently deceived, for the gull may well be classed with the crow and the English sparrow as a leader in independent thought and action in the feathered world. There are, in fact, few birds so accomplished as the gull.

It is, for instance, so completely a master of the air that it can fearlessly breast the fiercest hurricane. It is so completely at home on the waters that it can skim a living from their surface or use it as a cradle in which to be rocked to sleep. Again, it is so successful a land bird that it may go back among the farms and out-do the blackbird as it follows the plow and preys upon the grubs and worms that are turned up by it. It can keep warm among the icebergs of the North, or cool under tropic sun. It is at home on barren islands where the voice of man is never heard, or upon the docks of his busiest ports.

Among the mammals it is found that a beast of prey has much more intelligence than has one that feeds upon grass. It needs more intelligence, for it must outwit other animals that it may catch them as food. In the same way the flesh-eating birds are more intelligent than the grain-eaters. The kingfisher, for instance, must be wise in the way of minnows swimming beneath the water that it may be able to plunge from high above them and capture them. Doubly clever, however, must that bird be which takes its prize away from another bird that has already captured it, as the gull robs the pelican.

Another interesting fact that a study of birds develops is the superior intelligence of those that live together in colonies, over those that live solitary lives. The kingfisher, for instance, living alone on a limb beside the water, is wise in its way, but it knows but one trick, that of catching fish. The gull, clever and active as it is, gets its food in many ways. After it gets it there are a score of its own fellows always about that immediately try very energetically to take it away. They fight over every morsel, always noisy, greedy, jealous, hungry. The competition at home sharpens the sea gull's wits.

But the gull has many virtues, despite bad manners. Although it is careless as to what it eats and is largely a scavenger, it is spotless in its cleanliness. It keeps away from anything that would soil its dainty plumage. The grown bird is clad mostly in white, sometimes with exquisite rose color tintings, a suit which it changes but once a year, yet which is always clean. The best known gulls, the herring gulls, have slate-gray backs with black tails and wing tips. Those brownish-yellow speckled ones are the young birds, still in their pinafores, not getting grown-up clothes until they are three or four years old.

They are handsome, stately birds, these gulls, with wings so artistic in their proportions that they might have been designed by a great master. Every movement is full of grace; they are superb in flight, adapted perfectly into every mood of the sea, creatures of eternal beauty, probably as often placed on canvas as any living thing.

Wherever on this earth the land goes down to meet the sea, or any other large body of water, salt or fresh, there the gull is to be found. Whenever, in the Old World, an emigrant takes ship and sails away doubtingly into the unknown, it is the sea gull that pilots him forth and bids him the last goodbye. Wherever a traveler returns from many wanderings and approaches his native land, it is the gull that is the outrider, bringing the first welcome home.

The sea gull belongs to the order of swimming birds in the same way that the hawks belong to the birds of prey, the orioles to the perching birds that live in trees. When brought down to a narrower classification, it is found that the gull belongs to the "long wings," that is, in the tribe of swimming birds the gulls belong to the family of long wings. They are cousins to that greatest long wing of them all, the albatross, whose pinions may stretch eleven feet from tip to tip.

The gull, in fact, is of a group of birds that is rich in the development of feather covering, in which the plumage is almost unsurpassed either in beauty or usefulness. The feathers of the gull, the feathers of all the birds, are a unique structure in nature, are the peculiar property of birds, are to be found no place else in all creation except on birds. They are another of those commonplace things which few people come to understand, to know of what they are made, or how they came about. It might be worth while to stop for a look.

Feathers, strange to say, are but hairs in a different form. Hairs, in turn, are but scales long drawn out. All grow from the outer skin of animals and are made of the same material as that outer skin. They are but unusual developments of that outer skin. The outer skin is the part that peels off after one is blistered from his first swim in the summer sun. The claws of the tiger, the fins of the trout, the shell of a crawfish, the rattles of a snake, the horns of a bull, the teeth of a dog, the nails of your own fingers, the great horn on the nose of the rhinoceros, are but strangely developed hairs, are but feathers in different form.

One group of animals, the mammals, that group which feeds its young on milk, developed hair as a covering to protect its body and keep it warm. Another group, the fishes, developed its top skin into scales. Yet another, the birds, got started along a different line and the result was feathers. All worked with the same material, cuticle.

Scales undoubtedly came first, as fishes were long in the world before any animals took to the land. Among the early land animals that grew scales were the reptiles, particularly the snakes. The birds, a step further removed, still wear scales somewhat like those of the reptiles from whom they spring, on their legs.

The scaled animals were of the lower order, while those with hair and feathers were progressive, were higher orders. They were the animals that were advancing, were getting ahead. They were doing so by developing themselves, by learning to do new things. One group of the animals with coverings developed claws and learned to dig and make itself homes under ground. Another used its claws for the purpose of climbing and took to the trees. The birds were of this latter class.

When birds first developed this ambition to fly, their feathers were limber, fluffy, and quite ineffective. As the bird leaped from perch to perch, however, it kept flapping its wings and trying to make them help it. If you will watch a boy making a standing broad jump you will see that he does the same thing with his arms. The bird's feathers began to respond to the demand placed upon them. The quills in them stiffened. The barbs on the side of these quills began to grip the air instead of yield to it. Wing feathers became different from the other feathers. They had a job of their own, that of pushing against the air and helping the bird to jump through it.

A hundred thousand years of this sort of exercise

developed the specialized wing feather. It is light and stiff. Scores of barbs stick out stiffly on each side of it. Each of these barbs has hundreds of barbules, tiny hooks, which bind it to its fellows on each side. Pull apart the barbs on any flying bird's wing feathers and you can feel the hooks tearing loose. They can be plainly seen with a microscope.

So is a stiff, wind-tight feather developed, cupped on the under side to catch as much air as possible. It is bound in with its fellows, each covering the other like the shingles on the roof. Together they make a surprisingly light air paddle that meets with the greatest possible amount of resistance on the down-stroke and almost no resistance on the up-stroke.

The tail feathers of flying birds have stiff quills and stiff barbs because the tail is the bird's rudder and much used in flight. Other feathers serve special purposes. Those on the back of the bird must keep out the rain. A duck's breast must be water tight. On birds that do not fly, as, for instance, the ostrich, the feathers remain fluffy.

In general there are three kinds of feathers: the big ones that cover the body, part of which are developed into the wing and tail feathers; the downy ones that are common to young birds and exist as underfeathers throughout life on such birds as ducks to keep them warm; and the tiny, hair-like feathers such as those that make it necessary that a chicken should be singed after picking.

The wing feathers are seldom more highly developed than by the long wings, like the gull; the surface feathers seldom more delicately colored; no bird better uses its feather raft to keep it afloat.

The gull, on the whole, is a useful bird in its contacts with man. In the first place it is a scavenger. It eats up waste that might otherwise pollute the water. It follows all the ships that leave all ports until they are many miles out at sea, eating whatever may be thrown overboard. It is equally fond of the highly seasoned food of ships of Mediterranean, good French cooking, or sauerkraut and pig's knuckles from further north.

HERRING GULL PREPARING TO BREAK A CLAM BY DROPPING IT ON THE BEACH.

The gulls follow the fishing vessels and eat the waste when the catch is dressed. They loiter along the shore and devour anything from a shrimp to a whale that may be cast up by the waters.

They gather in all the harbors of the world and eat the waste. They are unofficial, coastline moppers-up.

In New York Harbor alone, by actual count, there are regularly at work 75,000 gulls. Such are their appetites that each is held to pick two pounds a day of waste from the water. The health authorities prize them as among their most efficient working squads. They believe there would be more illness in New York if the gulls went on strike. They are protected. He who shoots one must pay a heavy fine. They are coming to be protected in many places. The public in general is coming to recognize the sea gull as its friend. Some years ago there was a fad for sea gull plumage on hats and the money hunters threatened these birds with extinction. Public opinion has grown so strong against this sort of use of bird plumage, however, that it has largely been given up. The idea is growing that the man with a gun should not be allowed to kill whatever bird comes along without knowing whether it is friend or foe.

The sea gull often is a friend and helper of man and is seldom harmful to him. It even goes inland and lends a hand in the battle against the insect enemy. In Salt Lake City, Utah, there stands a monument to the sea gull consisting of a handsome marble shaft, crowned by a great ball upon which two of these birds are just alighting. It commemorates an event back in 1858 when the pioneers had just settled here and were growing one of their first crops. Down came hordes of black crickets from the mountains, and began busily to devour the crops. Starvation threatened. Then appeared the gulls in great numbers, bringing with them those appetites that are almost without an equal in the animal world, They devoured the grasshoppers and saved the settlement. They have rendered similar service on many occasions and at many points.

QUESTIONS

1. Tell the sea gull's scheme for eating clams. For stealing fish from the pelican. What idea of this bird's intelligence do you get from these tricks?
2. How would you go about observing gulls at close range? Describe the family life. Their combats.
3. Are gulls "gullible"? A versatile person is one who can turn his hand to many things. In what way does the gull show versatility?
4. Flesh-eating birds must be cleverer at getting a living than seed-eaters. Why? Are robber birds clever? Are those that live together in flocks?
5. The sea gull wears a spotless coat. Describe this coat. Describe the bird in flight. How does it manage to keep clean?
6. To what order of birds does the gull belong? To what narrower group? Is it widely scattered?
7. Feathers are peculiar to birds. The gull has a highly developed type of wing feather. Of what are feathers made? Give examples of other forms in which cuticle appears. Into what covering do mammals make cuticle?
8. From what lower type of animals did birds develop? Point out some of the characteristics of birds that show them to be the kin of reptiles.
9. How did birds learn to fly? Describe the development of the feathers through exercise.
10. Bring wing feathers, tail feathers, body feathers, to class, and show how each serves its purpose.
11. Show how the gull works as a scavenger along coasts and in harbors. What can you say of its appetite?
12. Is the gull a friend or a foe of the farmer? Why is there a monument to the gull in Salt Lake City?

CHAPTER III

THE MOCKING BIRD

THERE are many remarkable things in nature that are easily understood if one but gives them attention. There is, for instance, bird song.

Did you ever stop to realize that in all the world the only creatures that sing, with the single exception of man, are the birds?

The yowl of the cat, the bark of the dog, the chatter of the squirrel, the jabber of the monkey, the bray of a donkey, have no music in them, and cannot by any stretch of the imagination be set down as song.

The grasshopper and some of its cousins pull their shanks like fiddle bows across their wing cases and make rasping noises. The cicada, or harvest fly, has a drum head which it pushes in and out as one might the bottom of the dishpan, making a noisy but not musical clamor.

Voices that lend themselves to melody belong to the birds. As the fish was a pioneer in developing the backbone, as the frog family developed arms and legs, so was it the bird that first brought music into the world. And, as a matter of fact, it still comes near having a monopoly of it.

The human animal appeared, started in its path of progress, learned to think better than any other animal, and took note of the things round about it that were good.

It found that there had come into the world long before it this busy little body in feathers that had already developed the habit of perching on the top of a tree and pouring forth smooth volumes of pleasing sound. All along the road, since man's mind first began to take form, he has listened to that melody and found it sweet. Under the inspiration of birds man himself has slowly and with difficulty learned to sing. Today an occasional human being, after decades of training, may be capable of brief outbursts of pleasing song.

The mocking bird that was hatched last spring—a creature not so big as a man's hand—appears, perches outside one's window, and bursts into untaught song which surpasses anything that its masterful neighbor may, by all his study, produce.

Of all singers between the poles, the mocking bird carries in its throat a music box, the size of the tip of your little finger, that is more remarkable than any other music box of art or nature that ever came into being.

Not all the sounds made by birds, of course, are music. Their chatter is to them but their way of expressing themselves, of quarreling, scolding, or rejoicing. They have sounds with which they express different sensations, impressions, and ideas. Anybody who closely watches birds comes to know the meaning of at least part of their calls. There are cries of alarm, shrieks of pain, calls for food, as well as songs of joy. The mother mocking bird can ask her mate to do so practical a thing as bring more twigs for the nest she is building, while he can summon the whole bird population of the neighborhood to drive away a hated crow.

The notes of birds seem to harmonize more or less with their varied dispositions. The birds that plunder and steal generally make sounds that are unattractive. English sparrows chatter busily and obtrusively. Cardinals are as vivacious and gay in song as in appearance; the loon as melancholy and remote in voice as in habit; parrots, noisy and bad-mannered. The melody of the gloomy pine forest is more solemn than that of the orchard; night birds, such as owls, are gloomy and ghost-like, and the calls of some of them give one the creeps.

A short, sharp cry is a call of fright; an abrupt note of warning, continued and lengthened, reports that the coast is clear. The cry of pain is a low sound in the bird's throat. Even master singers, like the nightingale, shriek harshly when angered.

Bird calls are understood by all the creatures of the wilds. The plover of the Nile, for instance, is the watchman. It sits upon the back of the crocodile, and quarrels noisily with its fellows. Once let this sentinel give its short cry of alarm, however, and the water birds round about take flight and the crocodile dives beneath the water.

Among the birds the male is usually the noisy member. The real songsters are nearly all males. Males, among birds, also, as a rule, are the showy members of the family. In the great majority of cases they have the brilliant feathers and sit aloft and sing, while the female builds the nest. Sometimes, if her work calls for it, the female becomes a better talker than the male. The old hen in the barnyard, for instance, has the whole care of the young and has many handy words which she uses at her tasks that are unknown to the more showy but less busy husband. The breast of the little hen wild bird may swell with the joy of her brood, yet she cannot find the voice with which to express it. The male is the singer.

In many cases it seems that when Nature distributes her gifts she gives one bird a gorgeous suit of feathers, then seems to consider that she has done enough and denies that bird the gift of song. Often, on the contrary, when she bestows the gift of song she may dress the singer in a plain suit. The tropics abound in birds of gorgeous plumage and harsh voices. The woods farther north are full of plain, modest birds that would scarcely be noticed for their looks, but which are the world's sweetest singers. The mocking bird, the brown thrasher, the wood thrushes, all great song birds, are sober thicket dwellers, more often heard than seen.

The mocking bird is of the order of perching birds, those that live in trees and sit upon twigs. It is plain that such birds are quite different from those with web feet that swim in the water, those with long legs that walk and wade, or those that catch their prey in talons. Birds are put into orders based upon certain general characteristics and the mockers belong to the perching birds that live in trees. More than half the birds of the world, in fact, belong to this order, sit upon branches, and make insect catching and weed-seed eating their chief business.

Within the order of perching birds there are families, such as the orioles, the robins, and the wrens. The mocking bird belongs to a group that is closely related to the thrushes on one side and the wrens on the other. The catbird is a member of the same group, and, like the mocking bird, a great mimic. The thrasher, improperly called the brown thrush, though it is not a thrush at all, is another member of this family and a great singer. They are all lovers of thickets, are plain, unobtrusive birds that find their greatest glory in the beauty of their voices.

Down in the throat of every bird that has a voice there is a little box made of cartilage, like one's Adam's apple, and called a syrinx. It was long thought to be the bird's music box. The turkey buzzard, which is voiceless, has no syrinx, nor has the European stork, while all those birds with voices possess them. So it was held that this syrinx was the seat of the bird's song. Modern students of the voices of birds, however, have come to doubt that this syrinx can be the bird's music maker. They say it is too far down in their throats. It is, in fact, at the base of the windpipe, and it appears that the sounds that might be made in it would be muffled and silenced before they could get out. It has to do with the voice, undoubtedly, probably controlling the flow of breath from the lungs that makes singing possible.

The bird's music box, it now seems, is its mouth, the roof of which is a sounding board, together with the throat near it that swells out like a bellows, and the little gate between them called the glottis, with a narrow opening, which can be opened or closed to control the pitch. It is through this little slit that the bird whistles its songs. It was through this slit that it introduced music into the world. It was through this slit that it has caroled for a million years to sell to the world the thought that melodious sound is distinctly worth while, a beauty and a joy forever.

The mocking bird belongs to America, and does not exist on the eastern hemisphere. The sweet singer of Mexico and California is a little different from that of eastern United States. The most numerous race of mocking birds has its home from the Gulf Coast of the United States to latitudes as far north as the city of Washington, and is not migratory. One of the interesting tendencies of the mocking bird is an apparent effort to extend its range. A generation ago it was not to be heard in Washington, but was abundant fifty miles below that city, on the Potomac. Now it lives even beyond Washington. Its numbers are likely to increase in this new territory, if there are several warm years in succession. But an unusually cold winter may again kill most of the pioneers and give the advance movement of the race a setback. There is land hunger among the birds, however, as among men, and they keep striking out into new fields.

Mocking birds are nowhere more at home than in our southern states. The Dixie mockers are fond alike of the negro cabin with its fig tree and arbor of scuppernong grapes and the stately plantation halls with their beautiful grounds; and there they nest, close to that human companionship of which they seem so fond.

The mocking bird builds its nest in a way that indicates the carelessness of the great artist for details. It is placed in low bushes, or small trees, perhaps in the vine that runs over your veranda. Rough twigs of whatever material is at hand forms the base, while string, horsehair, bark ravelings, or cotton may smooth the inside of it for the young that are to come.

There is tragedy in the very lack of protection that

surrounds the mocking bird's nest. It is easy of access for the stealthy house cat that purrs on the mat during the day and appears the soul of innocence, but becomes a murderous demon by night. It often happens that such a cat eats up during the season the half-grown families of twenty pairs of singing mocking birds. She steals upon them in the night while they are yet helpless, and gorges herself. House cats, those that are still

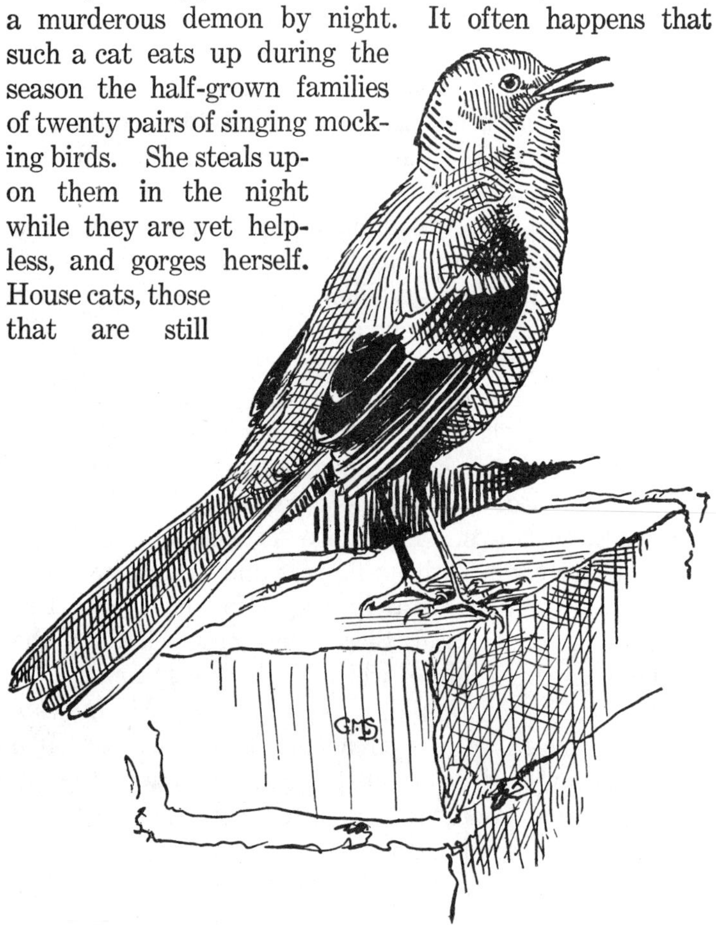

A MOCKING BIRD SINGING FROM A CHIMNEY TOP

tame, and those that have gone back to the wild state, are the worst enemies of the singing birds.

During the nest building the male mocking bird sings from the chimney top, from its perch in the thicket, from the limb of some tree in the orchard, or even on the wing, beginning with the dawn and keeping at it for a good eight-hour shift without ceasing, burstingly, leaping now and again into the air and turning what may be called a bird handspring in the fulness of his joy. Again, he works a shift in the afternoon, sings persistently in the twilight, and, like as not, will be found making the welkin ring if one happens to wake up when the moon is shining at two in the morning. By all of which he proves that this marvelous whistle of his has well-nigh the endurance of those made of tin and bought at the store for a nickel.

Beyond the range of the mocking bird to the North is to be found in abundance its cousin, the catbird, so closely kin, so alike in its manner of life, that the description of the life led by the one pretty well fits that of the other. It is unfortunate for the catbird that it was given a name so unattractive, merely because it happened, upon occasion, to mimic the "meow" of Tabby. No poet could sing the praises of a bird with a name so commonplace. Unfortunately, also, there is a bit of the roisterer in the catbird's nature, and it does not hold to the high plane for which the magnificence of its voice fits it. A dignified catbird with a creditable name like that borne by the nightingale would have been sung by all the poets.

The mocking bird may sing hidden in the thicket, unseen among the leaves of an orange tree, or in sight of all on the topmost bough of a forest tree. Its great masterpiece is the dropping song, its ultimate attempt to exhaust the last fiber of its being in melody, in one outpouring, the moment of greatest glory in song ever reached in all the bird world.

THE MASTER MUSICIAN

The dropping song, sung at the height of the mating season, is not often witnessed and has been described but seldom. It is opened on the tree top with an ecstatic trill, and continued with the wings extended and aflutter in the sunlight. Then comes phrase after phrase of quavering melody that sets the bird's whole being to palpitating. So moved does it become that it seems unable to retain its place on the tree top. It springs ecstatically into the air, some few feet, half falls, half flutters downward among the leaves of the tree, pouring out all the time such a wealth of sound as is heard only upon this, its greatest of days. Atremble with the madness of its effort, the bird clings indifferently to the branches, falls from one limb to another, drops by slow stages down through the tree, finally falls to the ground beneath it and lies there, limp and exhausted from the frenzy of its effort.

This is the mocking bird's own song, unaffected by those of its feathered fellows. The side of its nature from which it took its names reveals it in an entirely different rôle. Perched there in the thicket of an autumn afternoon, wide awake and attentive, it listens carefully to all that is going on about it, never making a sound. Then, as twilight comes on, it appears not in the rôle of grand opera star, but as a performer in vaudeville, full of pranks and humor.

During the afternoon this listener has heard the farmer's boy whistling to his dog as he started fishing, and now, the call is repeated. Dog Tray used to respond at once to this whistling, but now he is suspicious when it comes from the thicket. This mocking bird repeats the call of the young chicken lost from its mother, and puts that fussy person to clucking loudly. It caws as did the passing crow toward sunset, and screams like the ungreased pulley at the well. A nestful of young sparrows have clamored noisily when their mother appeared, worm in bill, and the mocker imitates the whole brood of them. He even mimics the notes of his own young ones, in their first efforts to sing. The twittering of the swallow, the song of the bluebird, or the song of the brown thrasher, one of its most ambitious rivals, is reproduced in its very detail.

The mocking bird does not stick to these songs of its fellows in slavish detail, but introduces into them certain variations, bars of its own. It is very likely to improve upon them, and is quite sure to give them better tone quality. It takes pride in singing down its rivals, in matching its voice against theirs, troubadouring from adjacent trees.

In America the mocking bird is generally admitted to be the best musician among the birds. In Europe the nightingale has won for itself the palm as prince of singers. This nightingale also is a shy creature of the thickets, has a closer relationship to the thrush, while the mocking bird tends to a stronger resemblance to the wren.

The nightingales come up from Africa and from the Holy Land in April and May, scatter all through Central Europe, and find a favorite home in England. There for two months they pour out their song in the twilight and the dawn. Master singers of the world their friends have called them, disregarding the claims of their rival in the new world. From the west comes the defiant claim that the mocking bird could listen awhile to the song of the nightingale and improve on it. Often has the possibility of matching these two singers of two hemispheres in song contest been considered, but it has never yet been worked out.

Emotional wild things fond of living in man's garden, mocking birds suffer cruelly when caged. Freedom is the breath of their lives; they belong in the trees. There they should be left to multiply and fill the air with music. There they will liberally pay their way in practical returns by devouring crop-destroying insects. They constantly fight on man's side in the great battle of the animal kingdom, that between man and insects. To be sure, they may eat a bit of fruit as it hangs ripe on the trees; such damage can easily be seen. The orchardman does not see and may not know that the insects the birds have destroyed would have done ten times as much damage while the leaves were tender and the fruit young.

Few people have ever stopped to think of the bird's place in the world as master of song. In the same way few people have ever stopped to think of the bird's proper place as game, as a thing to be shot when one ranges forth with a gun on one's shoulder. Certainly, when one comes to think of it, one should not shoot these sweet singers, which, in addition to the music they furnish, help man to keep the insects down that his crops may grow in greater abundance.

QUESTIONS

1. Name the different animals in the world that can sing. Which one invented song?
2. Birds have notes other than those for song. Have you ever heard any bird notes with a meaning you could make out? What are some of the messages these notes carry? Which sings the more, the male or the female?
3. In what part of the world do the best bird singers live? How do they dress?

4. Describe the peculiarities of perching birds. The mocking birds and their relatives belong to this group. What can you tell about the nature of these song birds?
5. What is the syrinx? Where is the bird's music box? How is it made?
6. Where is the habitat of the mocking bird? Where is it at its best?
7. Did you ever see a house cat catch a young song bird? Describe the cat's cruel methods of preying on such birds.
8. When does the male bird sing? Tell of the best bird song that you have ever heard.
9. Why is the mocking bird's cousin called a catbird?
10. Describe the dropping song.
11. The mocking bird plays pranks by imitating the sounds of other creatures. Name some of its impersonations?
12. What bird is the mocking bird's best-known rival? From what you have learned, which do you think is the master? Why do you think so?
13. Where do nightingales live? Where have you read of them?
14. Do mocking birds live happily in cages? What is your idea of the proper treatment of a mocking bird?
15. Where do mocking birds do harm? How do they help? Do they help most or harm most?
16. Why should we not shoot the mocking birds?

CHAPTER IV

THE MOURNING DOVE

WHEN mating time is on the coo of the mourning dove is to be heard more evenly distributed over a greater area of the United States than is the song of any other bird. It swells forth soothingly, melodiously, mournfully, from the Atlantic to the Pacific, from Canada to the Gulf, and on down the continent to Central America.

An interesting citizen is this mourning dove, or turtle dove as it is often called, when one sets it apart for a good look. This examination includes the pigeons, of course, for the pigeon is only a big dove, or a dove, a little brother to the pigeon. The two groups make up a bird order with traits different from all the rest.

Despite the fact that they are land birds, doves and pigeons have very short legs, yet they run about quite busily upon those legs. They feed almost exclusively on the seeds of plants, while nearly all the other birds feed more or less on insects and other living creatures. When drinking they keep their bills immersed, taking steady drafts, while other birds, as, for instance, the chickens, have to get a mouthful of water and lift their heads that it may run down their throats.

Doves are timid and shy, but at mating time especially

they fight among themselves, so that they seem not very fittingly selected as symbols of peace. They do not have the shrewdness of the crow, but nevertheless they are holding their place and increasing while other birds that would seem to be much more able and clever are disappearing from the face of the earth.

Domestic pigeons are only tamed doves in which certain peculiarities, such as fan tails, have been developed by selective breeding. Most of the varieties of tame pigeons come from the common rock dove of Europe. This rock dove is common and widely distributed and occupies a somewhat similar position on that continent to that of the mourning dove in America.

America supplies one of the most romantic and tragic of the world's stories of bird life, in which a kind of dove or pigeon played the leading rôle. When America was first settled and the pioneers pushed into the interior, they found there a certain dove which came to be known as the passenger pigeon. The passenger pigeon population of this area was so great that there is doubt if a single kind of bird in any other area in the world ever equaled it.

The passenger pigeon had one peculiar habit which was different from the manners of its cousins, the rock dove and the mourning dove. The rock dove and the mourning dove live in pairs quite to themselves through much of the year, and when they gather in flocks at certain seasons to migrate, these flocks seldom number more than two or three hundred.

The passenger pigeon, on the contrary, gathered in flocks so great that reports of their size are almost beyond belief. In the early days these flocks used not infrequently to pass overhead in such vast numbers that they would darken the sky for hours at a time.

Mourning doves do not crowd together to roost, but scatter out in the dry grass or leaves on the ground or even in small, loose flocks in the trees. The passenger pigeons, on the contrary, roosted in trees altogether and their huge flocks would settle down in such numbers in the woods that they would weight down large limbs and break them.

A reliable description of one of these great flocks of passenger pigeons in the early days tells of a wooded area

THE EXTINCT PASSENGER PIGEON, A PREY TO MAN'S SELFISH WASTEFULNESS

three miles wide and forty miles long which was their roosting place. The birds would scatter out while feeding by day for hundreds of miles, but would return regularly at dusk to this forest. The coming of these multitudes would make a noise like the roaring of the ocean on a rocky beach. So great were their numbers that they would pile one upon the other until the limbs would no longer hold their weight.

There was a noise of crashing limbs and of fanning wings of the disturbed birds throughout the night.

When the Middle West was settled, caravans of farmers would come from scores of miles around to slaughter the pigeons at their roosting places. They would kill them for fresh meat and also would salt them away for winter use. They would slaughter them in vast numbers, merely for the sport of killing. They would drive herds of hogs to the roosting places and fatten them on the bodies of dead doves. They would blind the birds with torches and kill them with sticks. They would burn sulphur beneath their roosts and suffocate them, causing them to fall to the ground in great heaps. They would shoot their guns through the masses of birds on the limbs and kill them, hundreds at a shot. The destruction was most cruel and wasteful.

With the coming of destructive men, the habit of the pigeons of gathering at roosting places proved fatal to them. As settlers grew more numerous they steadily reduced the numbers of the passenger pigeons. Finally it came to pass that there sat upon its perch in a cage in the Zoölogical Gardens in Cincinnati a single passenger pigeon, the only creature of its kind in all the world. All its unnumbered fellows had been killed. There it sat for some years, lonely and unmated, and finally, September 1, 1914, died of old age. With it passenger pigeons, a few decades earlier probably the most numerous single species of birds in the world, ceased to exist.

The mourning dove of America, which never assembles in such numbers as did the passenger pigeon, still survives. It is thought of, usually, as a plain, modest bird, but if one examines it closely one sees that it has many points of exquisite beauty. Look at the male, for instance. The upper part of this bird is grayish-olive brown, with just enough black spots on the wings to make the ground color striking. The top of the head is slate blue with a forehead of soft, wine-colored pink. The sides of the neck are tinged with bright pink that increases to a ruddy irridescence, just above the shoulders. The white chin fades into the wine-colored pink of the breast and this changes into a rich cream buff underneath the body. The blue-black spot on each side of the throat, the bluish sides, the olive-gray tail feathers, banded with black and tipped with white, the pink feet, the soft, beautiful eyes, all blend together in a harmony of quiet beauty that suits exactly the air of innocent shyness that seems a part of the very nature of this bird.

Throughout the northern half of the United States the mourning dove is a summer visitor; it comes in the early spring and stays well into September. In the southern states and on down through Mexico and Central America it spends the winter. For the four summer months it is everywhere the homemaker, absorbed with the rearing of its young. During this time it usually produces two families of two birds each, and brings them up in the way they should go. During this time each pair of mourning doves, living entirely to themselves, gives an example of devotion and happy home life that is so perfect that even among human beings it is generally regarded as a model.

The task of nest building belongs to the female, for the women usually do the work in birdland. The male dove is evidently not very critical of these structures or there would be less peace in the dove household, for the lady dove, be it known, is a very inefficient builder and housekeeper. She is, in the first place, quite careless as to where she puts her nest. Any tree crotch, any flat place

MOURNING DOVE IN COURTSHIP FLIGHT

on a limb, not too high, even the top of a fence rail or stone wall will do. In the plains areas they are quite content with the ground.

On any of these sites the mother dove piles a few twigs, whatever is available, with possibly some grass stems to soften them. They are so insecurely bound to the base that windstorms frequently upset them. They are so flat that it is a miracle that the mother is able to keep her two, plain white eggs on them until they are hatched, or the two clumsy, naked young birds within bounds until they have grown feathers. The nests are so easy for any crow, cat, snake, or other egg-eating and bird-eating enemy to reach that they are often despoiled.

But the dove husband does not quarrel with his wife on the score of her careless housekeeping. Instead, he seems quite proud of what she is doing. While she is doing it he indulges in the single bit of frivolity to which the dove lends himself in his whole career. He flies high into the air, spreads out his wings and tail in the most exaggerated fashion, and cuts many strange capers as he circles down the wind.

It is during this mating season, also, that the male wears all the bright colors in his feather trimming. It is during this time that he sings constantly his affectionate though to human ears plaintive song. With a throat full of sonorous melody he repeats through all the drowsy days of summer his soothing "coo-wee-oo, coo-o-o, coo-o-o, coo-o-o." When the nesting period is over, however, romance ends and the song dies until the following spring.

When, in the northern part of their range, the mating season for the doves is over, they begin gathering in groups of twenty, fifty, one hundred, three hundred which loiter about the stubble fields or the weed-covered open spaces. Then the weather begins to nip, and the doves are away to the South, flying at a speed that is by no means necessary to them as travelers, but which they have developed and hold in reserve for quite a different purpose.

The dove has no means of defense or escape from its enemies except flight. Its position among birds is something like that which the rabbit has among the mammals; in the face of danger its best resort is to show a clean pair of heels. It has wing power that is equaled by few birds. It has strength and speed in flight. The great development of the wing muscles, that is, of the breast, is what makes it so attractive an object of food to birds of prey.

This flight of birds, this mastery of the air that is so completely theirs, is one of the marvels of Nature, full of interest as to the bodily adaptations that make it possible. There is, for instance, the need of lightness for success in the air. According to natural selection theories those birds which were lightest succeeded best, were most able to escape their enemies. The heavy-boned, clumsy birds were eaten by the beasts of prey. In this way, without the bird knowing it at all, it tended constantly toward lighter bones.

Now the bones of birds are small compared with those of other animals. Where the bones of mammals are filled with marrow, which is heavy, those of birds are filled with lacework of pithy bone that is light or, in many cases, with nothing but air.

The lungs of birds which are not so highly developed as those of the mammals are lodged away in the rafters of their bodies, and are attached to the ribs. These lungs, however, are connected to air sacks that the bodies of mammals do not have. There are air sacs in different parts of the bodies of birds, between the organs, in their body cavities, even in their bones. They may draw on this reserve air when they need it, as, for instance, when they dive under water, or when they require an unusual volume of air for singing. It likewise plays an important part in flying, since it makes their bodies light.

From an engineering standpoint there are three principal ways of getting about in the world. When a man walks he puts his foot against the ground and pushes himself ahead. He gets along as does a boat that is being poled. The fish advances by wiggling its tail. It sculls its boat by working a single oar back and forth behind it. The bird uses the third method. It drives itself forward by pushing on the air on either side of it just as one drives a boat in the water by rowing with two oars. No bird gives a better example of this pushing against the air than does the dove. Watch it speeding just above the tree tops with its wings steadily flapping and you will see the perfect example of an air boat being rowed along.

The art of flight has been solved five separate times since the world began, and only five. The insects solved it first. They developed stiff, gauze-like wings, usually four in number, that worked on their sides a good deal as the cellar door works on its hinges. They usually buzzed these wings very fast, the bee, for instance, flapping them two thousand times in a second. The pterodactyls, or flying reptiles, probably were next in order. They learned to fly, lived a while on the earth, disappeared, but left imprints of their wings in fossil rocks. Then came the birds, with their forelegs turned into wings, and with feathers on them that would catch the air. They became more expert than most of the other creatures that learned to fly. The bat, a mammal covered with hair, which developed a great spread of bones of the hand with a membrane skin stretched over them, also learned to fly. Man, who not so long ago built a machine driven by an engine that carries him about in the air and glides rather than flies, has worked out the fifth solution of the flight problem.

The wing of the bird is its hand, very strangely developed to serve a particular purpose. Examine that wing carefully and you will find that it has, first, the single bone of your own upper arm, then it has, in the second joint, the two bones of your forearm. Finally, there is the wing tip, which has wrapped up in it a marvelous story if one cares to unravel it.

Examine the last joint of the chicken wing which you eat on Sunday, and you will find a small spur on the outside of that last joint. That spur is the thumb of this modified hand. Examine the small bones in the wing tip beyond this thumb and you will find that they represent what are fingers in other animals. This last joint of the bird's wing is a specially developed hand.

All the major land animals were once four footed. These four feet had their origin in the four fins on the sides of the fish, from which land animals came. Adventurous members out of the fish group practised coming ashore for some hundreds of thousands or millions of years, and in doing this learned to use their fins for climbing or walking on land. By exercise they developed four legs. All the mammals, birds, reptiles, and amphibians in the world today are descendants of this pioneer that developed legs.

In the beginning there was but one animal which had these legs. Thus all four-legged animals, if you go far enough back, have a common ancestor. The descendants of that ancestor struck out in different directions, lived under different conditions, and steadily grew less like each other.

For example, the seal, which was once a bear-like animal, living on the land, had well-developed feet upon which it walked, but the seal went back to the sea and its hand has become a flipper. Within the skin, nevertheless, are all the bones of a well-developed hand.

The bird evidently began with a good foreleg on which it walked, then got a habit of standing on two legs, and with the development of feathers, hit upon the marvelous scheme of using its hand as a wing. Then the chief problem was that of developing a frame which would make possible effective use of its stiffened pinions. The arm and not the fingers were found to do this best, so the arm grew strong and the fingers tended to disappear.

Thus, the specialized hand of the bird, with its supplement of feathers, became the highly effective wing which carries the bird into the air and on its long pilgrimages. For instance, it carries all the mourning doves that nest in the North back to the South when the cold begins to nip.

It also carries the homing pigeons, those most popular members of the dove family, on their errands of mercy as well as over long racing speedways, often at a speed of sixty miles an hour over a 75-mile course.

Doves eat few insects, so cannot be set down as allies of man in the warfare against this great menace. They eat wheat and corn, which is of value to man and are sometimes for that reason thought to be his enemies. Careful students of these doves, students such as those in the government's Biological Survey, however, state that little of the grain they eat is other than that left in the fields after harvest, which is waste.

Doves eat twelve months in the year, and their chief food is the seeds of grasses and weeds. In the stomach of a single dove were found six thousand foxtail seeds, and foxtail spoils pasture lands. In another stomach were seven thousand yellow sorrel seeds, in another five thousand hawk-weed seeds, and so on. Doves, year in and year out, are weed-seed eaters, and as such work constantly to keep down the same plants that man fights with hoe and plow. They are thus his good friends and allies and as such deserve good treatment.

Since the World War there has been little doubt in the minds of men that the carrier pigeon more than earned its daily ration of as much grain as it wished to eat. This specie of domestic dove is called more frequently homing pigeon, because of its unusal instinct for always returning to its home loft, no matter how far afield it may have flown or been taken.

During the war, homing pigeons carried messages of greatest importance between the trenches when no other means of communication was possible. In peace times they are trained by the United States Army Signal Corps for use in the air service for communicating in case of forced landings.

Much attention has been given of late years to the care and breeding of domestic pigeons of which there are 150 varieties. All render service to man in many ways. Besides being popular pets, the squabs or young pigeons of some varieties are considered a delicacy as a food. The homing pigeons are trained for the sport of pigeon racing, a sport that entertains a large group of people.

QUESTIONS

1 Name four peculiarities of doves. What is the relationship between doves and pigeons?
2. What do doves eat? Would you expect a bird that eats this food to be intelligent or simple? Why?
3. Tell the story of the passenger pigeon.
4. Tell the difference in the roosting habits of doves and passenger pigeons that may have saved the one and destroyed the other.
5. Point out some marks of delicate beauty in the dove.
6. What is the range of the mourning dove in its migrations? Where do they make their nests?
7. What sort of nest does the dove make? Which of the pair does the housekeeping? Feeds the young?
8. Describe the rôle of the father during the nesting period. The song he sings. Tell your observations of doves or pigeons.
9. What changes came in the mode of living of the doves? Their song? Their solitary habits? Where do they now live?
10. What is the chief aid to doves in escaping from enemies?
11. Birds, to be successful at flying, must be light. How are their bones made light? Describe the air sacs inside their bodies.
12. Does the bird pole, scull, or row itself through the air? Name the five creatures which, since the world began, have solved the problem of flight.
13. Explain how it is true that the wing of the bird is really a hand. Where, on the hand of a chicken, would you look for the thumb?
14. All vertebrates, animals with backbones, have four legs in some form. Where were legs first developed? How did the bird's fore leg become a wing? That of a bat? Why did the horse's hand grow as it did?
15. Is the dove helpful to man? What are some of the enemy plants they keep in control?

Chapter V

THE NIGHTHAWK

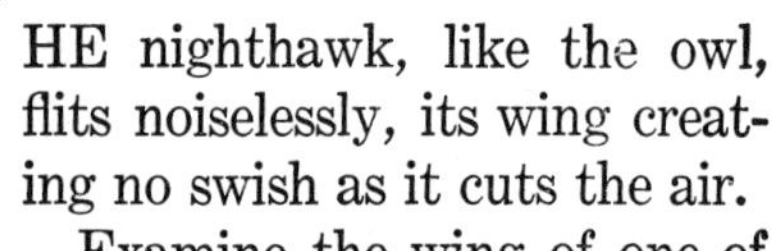

THE nighthawk, like the owl, flits noiselessly, its wing creating no swish as it cuts the air.

Examine the wing of one of these night prowlers and you will find that the feathers are not ribbed and harsh as are those of most birds, but have downy edges put there for no other purpose than to muffle their progress through the air.

Stroll along the old rail fence in June or climb up some rocky ledge and look for the nesting places of the nighthawks. Though their homes may be in reach, you are likely to miss them. This is because the bird is so camouflaged, because its color matches its surroundings so well as to make it practically invisible. It fades into the rail or the rock or the leaf-strewn earth so perfectly that it is nearly impossible to know where the bird ends and the background begins.

It affects protective coloring. It uses blends of black, brown, yellow, put on in mottled bars, thus causing it to merge with the objects among which it nests. It trusts to the fact that it is hard to see and will let you pass within a foot of it without moving.

It is such a flat bird, flat head, flat body, plastered

there so motionless on its nest that one can hardly tell it from a leaf. The woodcock also is artfully colored and has its nest on the ground, but it has such big, black, frightened eyes that one is likely to find it by first being attracted by those eyes. In hunting for nighthawks the best scheme is to look for their eyes. They seem to be aware of the danger of these eyes being seen, however, and narrow them like slits, thus doing away with part of the danger.

When you do find the nighthawk on its nest and come too near, it flutters at your feet as though wounded, thus inviting you to attempt to pick it up. This is a common trick of the ground birds, from the tiniest to the ostrich eight feet tall, when a stranger is near their nests. They want to distract his attention, to lead him away from this home he may destroy. Try to pick up this bird that seems to be wounded and it will flutter along the ground just out of your reach. When it has led you entirely away from its nest it will fly off as well as any bird.

The nighthawk has found a new solitude to take the place of the lonesome spaces that, thousands of years ago, before man began changing the face of the earth, used to be all about. And where are they? In the midst of big cities, where men are thickest, to be sure. This bird nests and rears its young in these city solitudes.

It often happens that the roofs of high city buildings are as quiet as the wastes of deserts. Even people who live in mere rows of three-story houses in the cities almost never go onto their roofs. Here the nighthawk is safe. Here it finds that it may lay its eggs and raise its young ones. Here it escapes most of its enemies as, for instance, the egg-eating snakes. Here it finds a place perfectly adapted to its style of housekeeping.

The nighthawk makes no nest at all. It lays its two eggs on the ground, on a flat rock, the flat side of a rail, the top of an old tree stump, the flat top of a building. There it sits on them until they hatch, there it feeds its young until the little birds have become big ones. It seems to get along quite well, thank you, and probably sneers at the stupidity of its feathered neighbors who take so much time with nest building.

The flat top of a skyscraper, tarred and graveled as so many of them are, just suits the nighthawk. There it is finding a new home. It sits there where the sun bakes down through the hottest of summer days and seems to like it.

The nighthawk is not a hawk at all and it does not usually fly at night. Its name, therefore, would seem to be a great deal of a misfit. It is more like the swallow than any other of the daylight birds. It is a swift, darting, insect-catcher like the swallow. It is often called "the swallow of the night." But in certain respects it looks like a hawk when it is on the wing, and hence its name. It may fly on a dark day or a bright night, but it is the twilight that it claims as its own. It is during the twilight hour, morning and evening, between day and dark, that it appears, flitting noiselessly as it gathers breakfast or supper.

Dusk is the time of insect flight. The bat, the only mammal which flies, knows this also and comes out to feast. The warty toad, there in your garden, with a bit of glue on the end of its tongue, snaps at insects in the twilight. The nighthawk, which some people call a bull-bat, and others speak of as the mosquito hawk, then comes out and banquets royally.

These nighthawks are migratory birds of the first rank. They winter in South America around the equator and far beyond, come up from the South in the early spring, and spread out like a great dragnet that embraces continents in its sweep. Slowly they move northward, foraging on insects night and morning as they advance.

They fly by night, feast in the twilight, and hide in the woods by day. As woods birds they are different from all others, except their first cousins, the whippoorwills, and the chuck-wills-widows, which can hardly be told from them. They do not alight on twigs, could not stand on a twig if they tried. In fact they cannot even stand up on the ground for more than an instant. They have strangely weak feet that they use very little. They cannot walk. They hardly use their legs at all.

With these weak legs, therefore, set far back on their bodies, they alight only on the larger limbs. They do not stand crosswise on those limbs as do other birds, but recline lengthwise. They stretch themselves along a limb, most of their weight supported by their breasts.

Most nighthawks do not begin to think of nesting until they get as far north as Tennessee. There the northern drift begins to thin out as one couple after another picks a place for egg laying. On goes the march with every day the numbers getting less. Some few go on as far as Canada, even as far as Hudson Bay, before nesting.

Here in the North the young are bred, and in September the drift to the South is begun. September, in middle latitudes, is the time to watch for the nighthawk hordes going south. They linger later in the southern states, feasting on the abundant supplies of insects of the autumn. Then they press on to Central America, Brazil, and Argentina.

One of the most interesting performances that is put on by the nighthawks may be seen by those who observe them as the idea of mating overtakes them in their northern march. As they pause for their picnicking along the way, two or three males may become rivals for the favor of a single lady nighthawk. She may sit demurely in her flattened pancake way on a tree stump in the clearing at the edge of the woods, and the males will put on their dare-devil stunts in airplaning for her entertainment. Each male will fly high into the air, possibly a thousand feet above her. Then he will suddenly turn eastward, will nosedive, will descend at a frightful rate of speed as might an airplane which had lost control. It will look as though he had every intention of dashing himself to pieces upon the ground. Just before he reaches it, however, he sets his wings and his tail, which is his rudder, in such a way as to put on all his brakes and throw all his pinions in resistance to the air.

When he does this, the wind among his pinions makes a most remarkable sound. This sound is usually spoken of as the booming of the nighthawk. So successful is this bird as a drummer with its wings, that the sound it makes of a quiet evening may be heard for a surprisingly long distance.

No sooner has one flight been ended by this crash of wings on air, than the bird turns about, again flies high, and again comes down to trumpet before the sweetheart he is wooing. Before long this damsel, by some method which many may not understand, indicates her choice of a mate. Thereupon, her other wooers go on their way to the North.

These birds of the twilight bear a very peculiar name in the Old World; they are called goatsuckers. This name goes back to early times when the shepherds held these flitting creatures of the twilight somewhat in superstitious awe. They came to believe that these strange birds visited their flocks at night and robbed the goats of their milk. They probably did follow the flocks and probably found that mosquitoes, themselves in search of food, buzzed about beneath the goats where the hair was thin, looking for a chance to bite. The twilight birds, trying to catch these mosquitoes, gave the shepherds a false impression. Thus this queer name of goatsucker, entirely undeserved, came to be applied to nighthawks and other members of their family.

The relatives of the nighthawks in the eastern United States are the whippoorwill and the chuck-wills-widow both of which names come from the peculiar calls which these birds repeat, through the still watches of the night, in a way that is somewhat awe inspiring. Their philosophy seems a harsh one, quite unsparing to poor William and to his bereaved mate even after he has departed from this world. The insistance that William be given the rod is likely to be reiterated hundreds of times between dusk and dawn by a single bird, as is the suggestion that his widow be put out.

There is a mellowness of tone in the call of the whippoorwill that is almost without a rival, though seemingly ghost-like and unreal in the dead of night. Sometimes at night it alights on the farmhouse roof and begins its hooting. The farmer, however, is unlikely to be disturbed by it and, being a practical person, has construed

A WHIPPOORWILL CHASING A BUTTERFLY

its coming in the spring to bring him a message. He holds that the bird does not say "whippoorwill," but "plant-that-corn."

These whippoorwills and chuck-wills-widows are shyer than the nighthawks, are less likely to be found about the abodes of men, are more inclined to hide themselves deep in the woods. The whippoorwill is much more widely known by its call than by its personal appearance. It is oftener heard than seen. If you should see it flitting about in the night, however, you would have much difficulty in telling it from its cousin, the nighthawk. It has the same flattened appearance and the same hawk-like expanse of wing; the same general color scheme, except that its black, brown, and yellow is put on in a different way. It wears a mottled coat, while that of the nighthawk is more given to bars. Its coat is brown where the nighthawk's is gray. The nighthawk, also, uses a device common to airplanes, of marking itself for identification. When its wings are spread out there may be seen on their undersides, just as on many air-planes, a distinct and clear-cut circle, done in white and standing out very clearly against the dark background of its wings. The whippoorwills have no such distinguishing mark.

The whippoorwills are more nearly true night birds than are the nighthawks. A nighthawk rarely ventures abroad after darkness is complete unless the moon is shining brightly. It likes semidarkness. It comes out for its hunting, in fact, on dark and cloudy days. It flies high. The whippoorwill, on the contrary, flies low, keeps going throughout the hours of darkness, and never appears by day.

Among birds of prey hawks work by day and owls assume the responsibility for the night shift, and go hunting birds and rodents while the day shift slumbers. Among small birds that live on insects the swallow is typical and hunts actively throughout the day. As the swallow goes home from work he meets the goatsucker shift coming on the job for the night.

As swallows are devourers of insects by day, so are goatsuckers active pursuers of these pests by night. It is the larger insects that are the favorite food of these nightbirds—the beetles, the moths, and the grasshoppers. The beetles, the moths, and the grasshoppers are man's worst enemies in the insect world, and so these quiet prowlers of the night are working always in the interest of man. They are helping keep down the hordes of insects that are seeking to devour his food supply.

The mouths of the goatsucker family are peculiarly adapted to this business of catching insects on the wing. When their flat heads open they have a surprising expanse of mouth. Their jaws hinge so far back that the mouth can open very wide. The purpose of this is that it may open a big gate for the entrance of the insect. To give it the effect of being even wider, the mouths of many of these birds have hair-like trimmings, bristles that extend out from them, that catch insects and guide them into the gate. With a net like this, one of these birds may fly into a swarm of mosquitoes, mouth open, and before it has gone far it will have captured dozens of them.

As the nighthawk has a cousin which tends a bit more to a love of the darkness than it does, so it has a cousin on the other side which tends more strongly to a love of light and which becomes a bird that hunts entirely by day. This cousin is the chimney swift, and students of bird life have quarreled for a hundred years as to whether it should be put in the family with the swallows or in that of the goatsuckers. Finally, basing their opinion upon the form of the body and feathers of the bird, scientists have pretty well come to an agreement that the chimney swift should go into the group which includes the nighthawks and the whippoorwills.

Like the nighthawks, this chimney swift has not allowed the increasing number of man creatures in this world to interfere with its comfort. As a matter of fact it has found that these newcomers provide for it an abundance of safe retreats in which it may nest and rear its young. They have built chimneys for all the world like the hollow trees which used to be favorite nesting places for swifts. To be sure, man's idea of what a chimney should be has of late been changing, much to the disapproval of the chimney swift. Men used to build great, wide-throated chimneys with room for wing play in them, chimneys with wide skylights that made it possible to see how to build nests far down in them. Now they are narrow and dark. Not so pleasant as of old, but still handy and abundant, and not often in use at the time the swifts come north.

By using these chimneys the swifts get along better where there are many men than where there are few. Even the meddlesome English sparrow that torments so many of the attractive birds, gives the chimney swift no cause for worry.

These swifts come up from the South as do the night-hawks, winging their way with great speed, and devouring endless numbers of insects as they come. They are given much to traveling in flocks, and even to frolicking in flocks after they have made their summer homes. In these flocks they play many prankish games there in the sky. It would seem that these birds have had military training, for they go through many of the evolutions of the parade ground. They form straight lines in the sky, dressing right and left, sweeping about, making military right and left turns. They form in columns in the sky, and go through fancy figures, wind snake-like as would an army marching down a crooked road. They have a game much like crack-the-whip, in which one end stops abruptly and the others struggle hard to keep from breaking the alignment. They play their games in the sky, and then, as though they had been dismissed, toward twilight, they dribble away into the chimneys round about and disappear.

The nest building there in the chimneys is peculiar. The chimney swift gathers twigs for its nest from the shrubs and trees while on the wing. It flutters past, seizing a twig, and breaks it off without alighting. In fact, a chimney swift never alights on the ground, or any other flat surface. It is one of the few birds that never touches earth except as the result of an accident. With the odd short legs it has it probably could not move about on the ground at all. It would be much more helpless on the ground even than a nighthawk, for it could not even lift its body to a standing position.

Its feet are most peculiar. The four toes all point

forward, and have claws on them with which they can grasp any such vertical surface as the brick inside a chimney or the wood inside a hollow tree. Having in-

CHIMNEY SWIFT BREAKING A TWIG

serted these claws as hooks, the chimney swift then uses its tail as a prop, a trick of the woodpecker. The tips of these feathers are turned into spikes to suit them to this purpose, and with the aid of tail and claws the swift thus hangs itself up to roost comfortably inside its chimney.

The twigs that it gathers while in flight it takes to the sooty solitude of its chimney and there shapes them into a nest. In making this nest it secretes a sort of glue with which it sticks these twigs together and with which it cements them to the chimney wall. With this combination of twig and glue it builds nests like a half-saucer. In the chimney this serves its purpose well, unless some irregular householder starts a fire in the summer time, smokes out the uninvited tenants of the chimney, melts the glue of their nests, and brings them crashing down into the fireplace.

Man, if he is wise, if he seeks an understanding of the nature of this family of birds, if he treats them in such a way as to serve his own best interest, will always spare the nighthawks and their relatives. If he kills one of them he is killing a friend. He is killing one who fights on his side in the battle against his worst enemy, the insect. He is acting as unwisely as would a soldier at the front who shot another soldier who was his ally. When a man is sent to war he is trained to know who is his enemy before he is given a gun. So is it coming to be held that he who goes forth to shoot the creatures of the air should know which are his friends and which are his foes before he begins his work of slaughter. Until he knows this he should not be allowed to go killing.

QUESTIONS

1. Why is the nest of the nighthawk so hard to find? How is its body camouflaged? Its eggs? How does it muffle its wings?
2. How do the ground birds try to lead one away from their nests? Tell of any bird that you have seen trying to do this.
3. How does it happen that nighthawks nest in cities?
4. What is the favorite time of day for nighthawks? Why? Name two other odd creatures that feast in the twilight.
5. Where do nighthawks go in winter? Describe their sweep to the North. Their return in the autumn.
6. What is peculiar about the legs of the nighthawk? How do they sit on limbs?
7. Describe the "booming" of the nighthawk. What is its purpose?
8. Why are these birds and their relatives called "goatsuckers"? What are the nearest relatives of the nighthawk in the United States? How did they get these odd names?
9. How would you tell a whippoorwill from a nighthawk? What are the differences in their habits?
10. What is the working arrangement that the goatsucker tribe seems to have with the swallows? In what respects are these birds similar?
11. What insects do nighthawks like the best? Why should we be glad that they eat them? How are the mouths of these birds especially built for trapping insects?
12. What bird is the nighthawk's daylight cousin? How does man provide nesting places for these birds?
13. Describe the games which the chimney swift plays in the sky. This bird also has queer feet. How does it use them in the chimney? What odd use does it make of its tail? Its glue?
14. Are nighthawks and their relatives friends or foes of man? Why? What rule would you make about shooting any kinds of birds?

CHAPTER VI

THE BOBOLINK

THE best-loved bird in New England is the bobolink, the rollicking songster of the meadows, a gay note in the landscape in its dress of black and white and yellow.

The most hated bird of the South Atlantic States is the rice bird, silent but for its oft-repeated note of "chink, chink," clad in its plain sparrow-like dress of brown and buff.

Yet, oddly, these two birds are one and the same. The rice bird is a bobolink in a different mood. The bobolink is a rice bird in its season of joy. Each mood has a costume to match. Each costume is accompanied appropriately by a voice to fit the part the wearer is then playing. Bobolinks, like Shakespeare's men and women, play various rôles.

Halfway between Massachusetts, where it is a bobolink, and South Carolina, where it is a rice bird, this traveler is known by yet another name. From Delaware to Virginia it is the reed bird, a game bird, and there is an open season through September and October when hunters go into the marshes and slaughter it by thousands.

And again, on its journey to South America, the bobolink appears in Jamaica, a British possession, under another name, and is known, because of the plumpness and fatness it has attained, as the butter bird.

These changes in name and manners are but high points in the remarkable career of this traveler of many titles and personalities. In attempting to follow the activities of a bobolink through the adventures of a single year one is puzzled as to where to start. Possibly as good a place as any would be in its dressing room when it is fitting itself out to play its rôle of joy in the North.

The bobolink hordes are on the move. They have banked up on the northern shores of South America and are changing their clothes. The female bobolinks are getting new suits, but they are such brown little nuns, always wearing dresses of the same pattern, that one would scarcely know they had been to the dressmakers. The males have been wearing these drab costumes also, looking for all the world like their mates there in the South, but now they are preparing for their day of glory. It is February and they are decking themselves out for that fling of madness that goes with their journey into Yankee land. When the molt is completed they step forth arrayed in that contrast of black and white and yellow that might shame the most fantastic Pierette, and are off.

Up by way of Central America, through the Caribbean to Florida, with a hop across the Gulf of Mexico or from island to island farther east, over the West Indies, straggling on through the states they come, during March and April, traveling chiefly by night.

On a sunny afternoon in the middle of May a New England farmer, driving home from the village, sees a flock of black and white birds mixed with brown ones fly up from their feeding grounds in the meadow and perch upon a tree, quite filling it. One starts a melodious if clamorous song. Others join in. Soon every male bird in the tree is bursting its throat with an energy that is almost fantastic. Then, by some mysterious signal, all stop suddenly and there is silence. It is a trick of the newly arrived bobolinks, back again to their summer home.

A BOBOLINK CAROLING IN THE MEADOW

Then the bobolinks pair off and find their homes in the meadows. During the mating season they are birds of the grass, of the open spaces. They make their nests among the growing stems on the ground. They make them loosely of grass, coarse on the outside and fine on the inside. They are of the same material that surrounds them, are part and parcel of it. So hard are they to see that, even in the bobolink country, few people have ever found them. They fade into the landscape.

And the modest mother bobolink! When she refused to change the color of her dress she knew that, in June, she would be hiding in these meadows and that with this gown, she would be well-nigh invisible. Bob might make himself conspicuous if he liked, but Roberta was thinking of the babies.

Yet Bob knows his business, for he always makes his clamor where the nest is not, thus diverting whomever might be searching for it. And when Roberta leaves her eggs for a brief few minutes, she creeps for a distance under the grass before she takes to wing.

Here in this safe nest the mother bird lays five or six eggs. No mere pair of youngsters for her as in the case of the mourning dove and the whippoorwill. She is a believer in big families. She has not come all this way for mere twins.

It is while Roberta is hatching the eggs and tending the children that Bob comes into his glory. There he sits in his wedding raiment on the top of a tree, or clings to a stalk of tall grass as a telephone lineman grips his pole, and pours out his soul. Or, maybe, he rises from among the buttercups and sings, lark-like, while in the air.

It is a tinkling song, full of madness, merriment, and melody. It is a loud, clear, rippling song, spilling over with recklessness and laughter. It is full of jerks and kinks and queer excursions. But it is the sweetest mad music in all the world, music that has inspired more love and poetry on the western hemisphere than that of any other caroler.

New England early took the bobolink for its own. Ever since the Pilgrims landed it has been a part of the life of this rock-ribbed section. It has been protected, encouraged, gloried in. It has brought to June in the meadows a rare charm, has become a part of the country life of New England.

This does not mean that the bobolink nests solely in those states. It is, to be sure, strictly an American bird, and one that nests only in the northern tier of states, and spills over into Canada. It is abundant throughout New York in summer, but little known during its joy season farther south. It spreads on to the end of space to the west. It likes the open, grassy country of the Dakotas, and is following the farmer and his meadows on through the Rockies, even to British Columbia. All this vast reach along the Canadian border is its breeding ground, its spring and early summer home. It is here that all the bobolinks in the world are hatched.

This singing, rollicking, strikingly dressed bobolink of the nesting season feeds almost exclusively on insects. That fact is more important to the bobolink than might appear on the surface. It is an important fact, because the bobolink is later to be haled into court and its very life may hang on its ability to prove a good character, a helpfulness to these men creatures who are to sit in judgment upon it. It is, in fact, an insect eater only during this season in the North, and almost exclusively a seed and grain eater when it later appears as the reed bird and finally as the hated rice bird.

Birds, to obtain protection from slaughter by man, must establish good reputations. If their habits are such that they render service to man they are likely to be treated kindly by him, to be protected by him. They will be set down by him as his friends. If, on the other hand, their habits are such that they are injurious to the interests of man, such that they, for instance, injure his

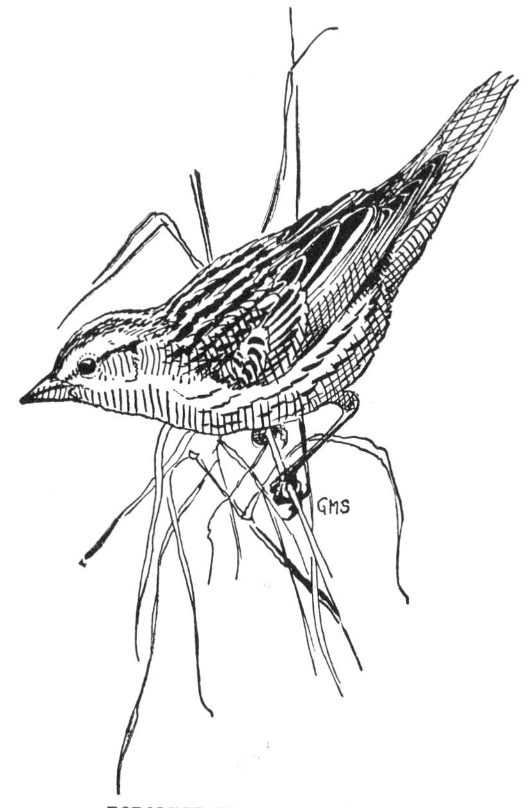

BOBOLINK IN WINTER PLUMAGE

crops, they are likely to be outlawed. They will be listed as his enemies and the sentence of death is likely to be pronounced against them.

This bobolink, singing so bravely from Maine to Vancouver, is an insect eater. It gathers, throughout the great area, a multitude of grasshoppers, beetles, cutworms, and caterpillars. It eats them, grows sturdy on them, feeds its lusty brood, hidden there in the grass nests, upon them. It works sturdily in the interest of man in his constant contest with the insect world.

By July the bobolink broods have grown their pin feathers and are taking to the wing. The family establishments in the meadows are being broken up. The old birds are shedding their feathers and the males are adopting that uniform sparrow-like costune of brown and yellow, are getting new clothes for the second time in the year. The proud and noisy songster is shedding his pied coat and making himself up to look exactly like his modest little mate who has been hidden there all summer beneath the grass. He is forgetting his mad and tinkling song and becoming a commonplace bird with but a mere "chink" between him and the ranks of those that are dumb.

Hardly recognizable in their new clothes these bobolinks begin to assume also a new manner of life, strange to their friends of the land of grass. Where they lived before as lone couples scattered through the fields, they now begin to flock together in groups. They begin to disappear from the meadows and to hunt the marsh lands. They are no longer meadow birds, but marsh birds. Overnight they have ceased to be insect-eating birds and have become seed-eating birds. The rank grasses of the marshes have come to maturity, have ripened their seed, and it is upon these that the transformed bobolinks feed.

Here in the guise of the reed bird the bobolink's relationship to its well-known cousin, the blackbird, is more apparent. The blackbird is fond of marshes, of traveling in great flocks, is a grain-eater, inflicting damage similar to that of the bobolink in its later rôle of the rice bird.

The bobolink is a member of the blackbird family. It must, therefore, admit kinship to the miserable cowbird which haunts the pastures where livestock graze and resorts to the most villainous trick in all nature, a trick peculiar to it in America and to the cuckoos in the old world. The cowbird is too lazy to build a nest and rear its young. Instead of doing so it lays its eggs in the nests of other birds, smaller birds like the warblers and other songsters, and thus starts a heartless tragedy on its way.

Watch the cowbirds among the herds. One of them feels that the time has come when she should lay an egg. She leaves the flock and begins hunting in the thickets where smaller birds are nesting. She knows that this is a low and heartless mission on which she is bound, for she slinks, like the sneak that she is, through the bushes. Soon she finds a little warbler's nest made ready for the eggs, probably with one or two of them already laid. Hurriedly she places her own egg in the warbler's nest and sneaks away never to return.

The warbler sits on this strange egg along with her own. It is supposed to be a part of the villainy of the scheme that the cowbird egg hatches a day or two quicker than do the warbler eggs. The unnatural guest in the nest has grown lustily before the little warblers hatch. It is naturally bigger and stronger than they are. It towers above them in the nest. When the parent birds bring worms it is the wide open mouth of the cowbird that always reaches above the others. It gets most of the worms. The tiny warblers sicken and die of starvation or are crowded out of the nest. Soon there is but one great, unnatural fledgling, thrice the size, in some cases, of the grown warblers that feed it, quite overflowing the nest. Finally, it is able to take wing and flies away to the pasture from which, in time, it will skulk back to plant tragedy in the nests of other little birds.

The migration of the bobolink back to the South begins early in the fall. And such a migration as it is! The route that the bobolink takes in its return to its winter home is most peculiar. All those hundreds of thousands of bobolinks that have scattered from Nova Scotia to British Columbia begin to gather into these seed-eating flocks, and to take wing. Strangely, they do not start directly south, to their winter home, but east. All the bobolinks of the interior of America come east until they reach the Atlantic seaboard. Those that have followed the farmer all the way out to Utah, for instance, come back to New Jersey before starting to their winter home. They have gone west by this route and they return the same way though they travel three thousand miles further in doing so. They have never learned the short cut south. The first great concentration in the migration of the bobolink mass that is to move every member of the tribe out of the United States is here on the Atlantic seaboard. Then on to the South. Oddly, these birds that flew mostly by night during the journey north, now fly by day as well, long lines of them overhead, hours in passing.

It is late in August and during September that New Jersey, Delaware, the shores of the Potomac, are athrong with reed birds, buffy little creatures that light in the marshes in such numbers, at times, as to cover ten-acre areas. This bird, drab, songless, mischief-working, is the glorious bobolink of the early summer, in its new rôle.

The months of September and October along the Atlantic seaboard are the open season for shooting reed birds. The shooting is best when the flocks first descend into the marshes. They are mostly young birds, untaught as to the practices of man and consequently unafraid. They light in masses among the reeds and proceed to fill their crops with the seeds that in their abundance are weighting down the heads of these plants. The wise hunter appears at sunset when the birds are gathering in dense masses to roost. He shoots through this mass and the birds arise from the reeds in huge flocks. Ignorant of the danger that awaits them they circle about in flocks, and each time they pass near a hunter he fires his shotgun through them, killing scores at a shot.

The slaughter of reed birds in the marshes is always very great when the season first opens, for the birds have not learned to be afraid. Before long, however, their ranks are thinned, the birds become shy, and shooting ceases to yield such big returns.

Thus passes the chapter in the life of the bobolink during which its masses are concentrated along the Atlantic seaboard and it is regarded as a game bird.

Then opens another chapter. On the night of August

twenty-first, according to a ten-year record, the advance guards of the reed birds enter South Carolina and become rice birds. The cultivation of rice in South Carolina and on down the coast to Florida is a considerable agricultural industry. At the season of the year when the rice birds begin to arrive in their huge flocks in the drift to the South this rice is "in the milk." This means that the kernels have formed, but have not yet hardened into the grain that will later be harvested. The grains are still soft.

Rice in this condition is a favorite food for these birds that were bobolinks in New England and reed birds in New Jersey. They descend upon the rice fields, perch upon the stalks, and squeeze out the milky kernels in unlimited quantities. The managers of the rice plantations muster great forces to assail these swarms of birds and attempt to drive them off. Among the rice lands the bombardment of these visitors can be heard from dawn till dusk and the slaughter is stupendous. So numerous are the flocks, however, that it seems to have little effect upon them and great damage is done to the crops.

It is because of this damage that the rice bird is the plague in this part of the South, and is more hated than any other migrant. It is because of the most remarkable change in its nature that the dearly loved insect-eating song bird of the North becomes the rice-eating menace and outlaw of the South.

Because of this change in nature, North and South, the bobolink has become the center of a controversy that has lasted through many years. Bird lovers in the northern states, where the bobolink is so attractive, protest against the slaughter of their favorite as it appears in the guise of the reed bird and the rice bird all down the Atlantic coast. They maintain that during the greater part of its residence in the United States it is an insect catcher, most helpful to man, and that it is the prince of meadow song birds. They maintain that the good that it does during this part of its career and the joy that it brings in its nesting season, more than counteracts the damage that it does to the rice fields of the South. They maintain that rice in the Carolinas is at best an unimportant crop and point to the fact that the great rice areas of Louisiana and Texas are off the bobolink route and remain undisturbed.

The rice growers respond by saying that they know this bird only as a destroyer of crops. It catches no insects for them, sings them no songs. To them it is only the destroyer. Why should they shoulder this heavy loss that people a thousand miles north, strangers to them, should benefit? Their loss is a concrete one, measured in dollars and cents, and runs into hundreds of thousands every year.

New England is complaining that the bobolinks are growing scarce up there. Around New York City where they used to be plentiful they can now hardly be found. The cause of this is attributed to the open season for shooting reed birds. Those government authorities on birds who work in the Department of Agriculture do not believe bobolinks as a race are growing scarcer. There still seem to be as many of them as ever going through the Florida funnel during the migration time. They hold that the bird is changing its range, that it is finding a new home in the West. It is following the farmer westward and is thus being weaned away from those states which were once its favorite abode. Birds sometimes do strange things of this sort, sometimes cease to exist in regions where they were once plentiful and become plentiful where they formerly did not exist.

The authorities of the Federal Government at Washington and of the various states along the Atlantic seaboard have listened to the plea of the rice growers and to the plea of the hunters in the states south of New York and have established an open season during which reed birds and rice birds may be shot. That season in most of these states spans the months of September and October. Thus does the bobolink become the only migratory song bird in the entire list that is officially declared a game bird and that is extensively killed by hunters.

The drift of the bobolinks from the West to the Atlantic seaboard and down a narrow strip along that seaboard is one of the most remarkable concentrations of birds in all the story of migration. The "bobolink route," in the great movement of birds from the North to the South, is one of the most definitely outlined courses traveled by any great family of birds. After despoiling the rice fields these birds follow on down the peninsula of Florida. From Florida they hop across to Cuba, from Cuba to Jamaica, and from Jamaica they make the four-hundred-mile jump to the mainland of South America. This is the "bobolink route."

At the season of the year when bobolinks migrate there are vast numbers of other birds traveling south.

To be sure more of them cross the Gulf of Mexico from Louisiana to Yucatan than travel any other route, but this journey of the bobolinks is by no means unpopular. When they take wing from Cuba they make up the great mass of the passengers of the air, but they are not without traveling companions. Going along with them is the trim little Alice thrush which made her summer home in Quebec. With them also is the New England cuckoo that has been sounding his melodious call through the quiet summer. There is the nighthawk, silent huntsman of the dusk, and the bank swallow that raised its young in Labrador. The tanager, a turncoat, like the bobolink, has given up his scarlet and black livery for one of yellow-green, and is now traveling south, with the black poll warbler that has journeyed all the way to Alaska. There is the vireo, the king bird, the wood thrush, and many another bird that is a transient summer visitor to the North. All are covering this water waste in a journey back to tropical winter homes in Venezuela or along the mighty Amazon, or in the forests that fringe the Andes.

Among all these, few are more ambitious in the flights they take than the bobolink. These birds pause but a little while in the northern part of South America. They span the great jungle forests of the middle part of that continent, go on far beyond the Equator to where southern Brazil merges into the waving, grassy pampas, and there among the marshes that border the Rio de la Plata they find the solitude and the seed-bearing plants that offer them an abundance for the five months that they spend in the south, the five quiet months in their otherwise eventful lives.

Thus is the cycle of their year brought around again to February, when the time has come for putting on striking raiment, tuning up their voices for their concert tour, and starting again to their northern home, in which they will play a rôle quite different from that which is theirs throughout the greater part of the year.

QUESTIONS

1. The bobolink changes its make-up and seems an entirely different bird in different places. What is it called in New Jersey? In South Carolina? In Jamaica?
2. Describe the change of suits at the time of the start north. The difference in suits of the male and female. The trip north.
3. Where do bobolinks nest? How does the mother bird employ camouflage? How many young ones does she hatch?
4. How does the father bird employ himself during the nesting season? Describe his song.
5. Point out on the map the area occupied by the bobolink in summer.
6. What does the bobolink eat during the summer months? How is this diet helpful to man?
7. In July the young birds are ready to fly. Explain the change that comes over the parents; the difference in habit which they take on.
8. They have now become reed birds. What do they eat? What is their song?
9. To what well-known birds are bobolinks related? Tell the story of the cowbird, the sneak and villain of birdland.
10. Describe the concentration of the bobolink flocks at the end of summer. Why do they come east to the Atlantic?
11. When and where may the reed birds be found? How do they differ in appearance from the bobolink? How are they hunted?
12. Show how the reed bird becomes a rice bird. What harm do they do? Compare the feeling that the rice planters have for them with the regard of the New Englander for the bobolink. Do you believe, from the evidence, that hunters should be allowed to shoot reed birds?
13. Trace the "bobolink route" back to South America. What are some of the bobolink's traveling companions on that route?
14. What sort of place is the winter home of this changeful bird?

CHAPTER VII

THE COMMON CROW

THE crow, it would seem, is one of the smartest of birds, and has a happier knack of learning man-like tricks than any of the others.

There is, for instance, the true story of the pet crow that got to playing with the top of a baking powder can, crows being fond of bright objects. It accidently released this lid on the inclined flat board at the side of the kitchen steps. The tin slid down the board. The bird, noticing this, brought it back to the top, and let it slide down again. It had learned a trick which pleased it and which it repeated without end.

Finally, by accident, the crow stepped on the tin as it started to slide. He went down with it. In doing so he learned how to toboggan, a sport in which he thereafter frequently indulged.

An odd thing about pet crows is their tendency to become great friends and playfellows with the dogs of the household. In their association with these dogs, however, other characteristics of the crow, its seeming possession of a sense of humor, its tendency to indulge in many pranks, are likely to assert themselves at the dog's expense.

A tale is on record of a crow which had a dog playfellow, a dog which had been trained to run after sticks which its master threw, and bring them back. The crow, having observed this, would wait until the dog was sound asleep, then get a tempting stick, nip the dog's toes, and wake him up. When he was aroused the crow would seize the stick and start to fly away with it. The dog would conceive it his duty to retrieve the stick and would start after it. The crow, however, would fly away just out of the dog's reach, and take him for a wild race across the pasture.

It is an instinct with crows to pull up young plants, that they may devour the seeds from which they spring. Pet crows, for this reason, are likely to be troublesome about the garden and to require discipline. They soon learn that pulling up these plants is forbidden.

One lady with a pet crow tells how she worked one afternoon among her young aster plants in the garden. She had dug out the weeds and the trash in this aster bed and laid them in neat piles. The telephone bell rang in the house and she went in to answer it. When she returned she found that the crow had pulled up all her aster plants and piled them as neatly as she had piled the weeds.

In another similar case there were geranium plants on a balcony which had been forbidden to the pet crow. One afternoon this crow was observed playing happily in the garden for hours. Finally, he seemed to get an idea, skulked quietly into the house, up the stairs, and out on the veranda where the geranium plants were growing. He slyly pulled them all up, piled them neatly, and stealthily returned to the garden.

Then there is the crow's favorite prank, that of annoying the washerwoman. When the wash is nicely on the line at eleven o'clock on a Monday morning and the clothes-pins are firmly in place, the pet family crow likes nothing better than to steal along the line and pull them all out again.

The intelligence for doing this sort of thing is very unusual in a bird. It is possessed by very few birds out-

MISCHIEVOUS CROW PULLING CLOTHES PINS OUT

side the members of the crow family, ravens, jay, and the like. Members of this family apply their higher intelligence to many of the everyday actions of their wild life. Either as pets or in the wild state they have a peculiar habit of gathering and hoarding bright objects, these seeming to have a strong appeal to them. They will steal the bright toys of children, glass bottle stoppers, spoons, and other shiny objects, and store them away. Wild crows have been seen even to dig garnets out of soft stone for their hoards, to visit camping places for bits of tin, to forage regularly for treasures that they may store away.

They apply their unusual intelligence also to the business of keeping alive and prospering. They match their wits constantly with those of man who has long considered them an enemy. Though their chief home in the western hemisphere is the well-developed farming region of central United States, and despite the fact that man seeks their destruction and death, they live and prosper.

As the flock of crows forages in the autumn corn field, for instance, eating what the harvesters have left or raiding the shocks, one sentinel sits aloft with an eye out for the actions of this man creature. A boy appears, whistling on his way to school. The sentinel pays no attention to him. The hired man brings his team into the field for the plowing. Still the watcher is undisturbed. The farmer appears with a gun on his shoulder. The sentinel crow sounds an alarm that starts all his fellows for half a mile around scurrying to safety. The farmer must use much cunning to get a shot at a crow.

The crow is styled the Robin Hood of the bird world. He is a bold, audacious, destructive robber. What he wants he takes. He takes it by stealth and cunning and lives with a price on his head, noisy, brazen, and swashbuckling, but none the less a great deal in terror all the time. He and his cousin, the raven, are the giants among perching birds, the largest of them all. He is a magnificent flier, is at home in the trees, is an active walker and runner, quite superb and stately in the manner of his gait on foot. He firmly believes that the world belongs to him and sets out to possess it.

He is a husky fellow this crow, nineteen inches long, strong of limb and beak. He ranges around the entire world. Like the flag of the British Empire, the sun never sets on him. There are many varieties of him, fish crows, carrion crows, hooded crows, but all have the same basic characteristics. Wherever he is found he is a clamorous, self-assertive, pugnacious creature, suspicious but curious, bold and at the same time shy, the possessor of an unlimited appetite which turns itself to many varieties of food.

The crow is the most widely distributed and the best-known bird in the world. Among its relatives are the jays, magpies, ravens, rooks, and jackdaws, all birds differing a great deal in appearance and habits, but all having the same aggressive, self-assertive, and meddlesome characters, the same all-devouring appetites, the same mischief-making dispositions.

Crows delight in the discovery of the great horned owl, alone in the twilight, and take much pleasure in pursuing and tormenting him. They persecute hawks and eagles in the same way, and those fierce birds of prey flee as though in danger of their lives. These same crows set upon bald-headed vultures, resting on the limbs of some dead tree, but vultures, wise in their way, pay no attention, for they have discovered that the crow's attack is only bluff, and that there is no harm in it.

One of the most remarkable phases of the crow's life in the wild state is the public meetings that they have often been seen to hold, meetings which appear to be trials, and

CROW TORMENTING AN OWL

are reported to result in executions. Clamorous flocks of crows now and again are seen to gather in some wood and begin making a vast amount of noise. There seems to be a storm center in the middle of these flocks, and after much deliberation it resolves itself into what appears to be an execution squad of six or eight members. This squad sets upon some individual crow which seems to have been condemned, and beats it to death. The mystery of whether this victim is a criminal, a religious sacrifice, or the victim of some strange rite of the bird world beyond the capacity of man to understand, may never be solved.

The greatest gatherings of crows, however, are at their roosting places. These are to be found in autumn and winter. In the spring they have scattered far and wide, have broken up into individual families as is the way of birds. Each pair has built a nest. More than likely they have looked down on the world from the top of some lonesome pine. The lookout posts near the top of the masts of ships are called "crow's nests" because they look like the homes of these birds.

Here they raise their broods of from three to seven young ones. As soon as they are old enough to fly, the instinct of sociability begins to assert itself among the crows. They keep together in family groups. These join with other family groups. Crow parties for foraging purposes are likely to consist of scores or hundreds of individuals.

Crows do little migrating. They may nest a bit farther north in the summer than they roost in the winter, but they do not desert the middle latitudes just because the thermometer sometimes flirts with zero, and Mother Earth tucks herself for a while in a blanket of snow.

The great central roosting places of the crows are usually in about the latitude of Washington, D. C. For many years there was a crows' roost just across the Potomac from Washington, and in it were assembled every night not less than two hundred thousand birds. Near St. Louis there is usually to be found such a roost with nearly one hundred thousand lodgers. There is another famous roost near Peru, Nebraska, and a number in Oklahoma. These roosts are to be found here and there throughout the middle United States, and east nearly to the coast, the wooded areas along the Delaware River being a particularly favorable resort.

A visit to one of these roosts is an interesting experience. An hour before sunset groups of crows that have spread out for possibly one hundred miles to feed begin winging their way homeward. Many of them arrive before bedtime and gather in groups on near-by fences, in fields, and in trees, there to visit and gossip and clamor and romp as might so many children in advance of the ringing of the school bell. Then, as night comes on, they group themselves in the trees that are their regular abode and settle down for slumber, a sleep that may be unbroken unless that horned owl which they teased by day sees fit to take his vengeance during the darkness when he has so great an advantage over his daylight tormentors.

Whether or not the meetings of the crows that are followed by executions are trial courts, it is true that the government of the United States has such a court before which it has arraigned and tried the whole crow tribe for its life.

It is not generally known that the government has a bird court before which many of the feather-wearers have been haled for trial. It takes testimony, weighs evidence, passes judgment, sometimes the death sentence.

The federal bird court is the Biological Survey of the Department of Agriculture, the business of which is to study birds and beasts in their relation to man. If it finds that a bird or a beast conducts itself in such a way as to be injurious to the interests of man, it condemns it, and may recommend its execution. If it finds that such a creature has been falsely accused, it publishes the findings and clears the defendant.

This federal court collects evidence with great care and its findings are almost beyond question. But the manner of these trials is unique.

The mild mourning dove, for instance, was once haled before the bird court and accused of being a despoiler of the wheat fields. It was urged that the dove ate so much wheat that she greatly injured the farmer. Her accusers claimed that she should be wiped out.

The bird court set out painstakingly to get the facts. Whether the dove injured or aided man, it argued, depended upon what it ate. The task in hand was to determine the bill of fare of the dove.

Scores of doves were killed in scores of states and at scores of different times of the year. The killing was done by members of the field force of the Biological Survey. The stomachs of all of these doves, each with a record attached, were sent to the laboratories in Washington. A scientific examination was made to find out just what was in each stomach. The record was set down. So was the fact established as to just what each of hundreds of doves had been eating. So was it determined that doves ate wheat hardly at all until after harvest, at which time it was waste wheat picked from the ground. Thus they did man almost no injury. It was shown that, the year around, the greater part of their food was weed seed. These weeds crowd out and injure cultivated crops. The dove was working valiantly in keeping the weeds down. It was helping man. It was acquitted by the bird court, was given a certificate of character, and was commended for its helpfulness.

The crow was haled into this bird court. One of the charges in the complaint was that this individual was a trespasser in the corn fields and a destroyer of the crop. Direct evidence was produced which showed that at the time corn was sprouting this crow went up and down the rows, pulled up the corn, and devoured the swollen kernel from which it came. This softened corn was a favorite viand of the crow and its young.

There was a second complaint that the crow was a marauder about the barn yard. It was particularly fond of eggs laid by hens and turkeys. It would visit their nests, sink its bill deep into an egg and fly away with it, later to gulp its contents. It was shown that a single crow had thus carried away sixteen eggs laid by a turkey that had stolen away to make its nest. The crow, it was shown, was a thief of, and destroyer of, eggs. It should be punished as such.

Yet other witnesses appeared to swear that the crow was guilty of even greater offenses. It had been caught in the act of breaking up the nests of the smaller birds, birds that were helpful to man, that made music for him. It would visit the nest of a thrush, for instance, and gobble all its eggs. It had been seen to approach the nest of a mocking bird, while its owners screamed and pecked, and had eaten the five little birds as though they had been so many grasshoppers. It was a vile murderer of baby song birds, a destroyer of their homes, and it should be punished as such.

Character witnesses were produced to show that this bird was a recognized outlaw, that its reputation was bad. Its very melancholy appearance caused it to be looked upon as an evil omen. It was generally known to be a loud-mouthed roisterer. It would rather steal its food than come by it honestly. It was quiet enough when it crept into the cherry tree for fruit, but noisy when after discovery it flew away. It ate carrion, often the flesh of animals that had died of disease, and was accused of carrying the disease to other pastures.

The prosecution here rested its case. It had undoubtedly made out a very strong case. It asked that sentence of death should be passed upon this destroyer, thief, murderer, and that every man with a gun be asked to become an executioner.

The government tested the accusations of the prosecution and compiled the evidence for the defense. It did this by that strange method of analyzing the contents of stomachs. It would find out what the crow ate that was harmful or was helpful to the interests of man. It would gather in crow stomachs from everywhere, at all seasons, and would find out what they contained.

. The field force began sending in crow stomachs. They kept at it until they had sent in more than two thousand of them, stomachs from all sections, taken at many seasons. An expert was kept at work very carefully analyzing these stomachs for five long years. The exact facts were worked out as to just what the crow ate that was against the interests of man and what it ate that was beneficial to man. Finally the balance was struck.

It was found that corn is the staff of life of the crow. It eats more corn than anything else. It eats it at planting time which is a great annoyance to the farmer, but the harm it does can be mended by replanting. It eats it at the roasting ear stage, causing much to spoil that it does not devour. It steals corn from the shock. It eats much corn that is left in the field and otherwise wasted. It is guilty of material injury to the corn crop. In places this amounts to destruction.

But just here was found some important evidence for the defense. In the stomachs of these crows were found many insects. They eat many May beetles and the white grubs from which they grow, and these are crop destroyers. They eat great quantities of grasshoppers and there is always a race between grasshoppers and grass to see which will survive. Caterpillars are the great leaf eaters, and it was found that these mushy fellows were a favorite baby food of the crows. They feed their youngsters great quantities of them. Every little crow eats three times its weight in caterpillars during the time it is in the nest.

Thus the question arises as to whether the insect eating of the crow does not do the farmer as much good as its corn stealing does harm. It is estimated that it probably does, but, unfortunately, while the benefits are general, the losses all fall on the corn farmer. Corn farmers are justified in killing crows by gun or poison wherever they can.

The stomachs showed that crows occasionally eat eggs and chicks from the barn yard. They do eat the eggs and young of song birds, birds that are themselves insect eaters. This is a loss to man.

Set over against this is the fact, proved by the stomachs, that the crow is an enemy of rodents, eats many small mice and rats, and many young rabbits. Here, again, is a service to man, for the rodents do much damage.

There are certain circumstances under which the crow does great local damage. There was, for instance, the case of the almond orchards in the state of Washington. The crows found that almonds at a certain stage of development were very much to their liking. So they came into the orchards in great flocks, knocked down the nuts, flapped to the ground and devoured them. Thus would they strip whole orchards.

The crow invaders can only be met by cunning, and these almond growers worked out plans for driving them away. They poisoned fallen almonds beneath the trees. Occasionally a crow got this poison and was either killed or made very ill. The others took note of these casualities, maybe the farmers took pains to exhibit the dead birds; at any rate, the wise crows came to realize that secret danger lurked in these orchards and so avoided them and let the almonds alone.

The conclusion of this bird court of the government is that, on the whole, the crow does a great deal of harm. This harm is about evenly balanced by the service it renders. It is part friend, part foe of man. The government will make no protest if farmers take steps to decrease the number of crows. The government will even go so far as to give advice as to methods of poisoning and trapping crows. Yet it is not greatly distressed over the crow as a menace as it is in the case, for instance, of the house rat. On the whole, if man must have a bit of target practice, it will do no harm to take a crack now and then at an old black crow so that his race may not become too numerous.

QUESTIONS

1. Crows, being intelligent birds, learn tricks. Tell of the crow that learned to toboggan. Of the trick one played on a dog. On the washerwoman. Do you know any other crow stories?
2. Crows know that man considers them enemies. Show how they keep watch for him.
3. Give a general description of the crow, his nature, ability, size, appearance, home. What are his relatives? What are their relations to the birds of prey?
4. Describe the public meetings of the crows. Their roosting places. Their nests. How many young ones are there to the nest? How do their habits change after the youngsters leave the nest?
5. Where are the winter homes of the crow? Is there a crows' roost near your home? It would be a stirring experience to visit it. Describe what you would expect to see. At what season of the year have you seen most of the crow?
6. What bureau of the government studies the habits of birds? How does it find out if a bird is helpful or harmful? What did it find out about the dove?
7. What were the charges in the bird court against the crow? How long did the trial last? What crop was it found to injure most? Show how it does some good.

8. Should corn farmers kill crows? Should bird lovers kill them?
9. How did the crows prove harmful in the almond orchards? How were they driven away?
10. Altogether, is the crow a friend or enemy of man? Since the balance is so close might we not agree that here is one issue about which we need not worry? Personally, would you shoot a crow if you got a chance?
11. Which do you think is more intelligent, the crow or the sea gull? Why? What other bird do you know that you think is as wise as the crow? Why?

CHAPTER VIII

THE PENGUIN

HERE is a peculiar bird, in some respects more like man than any other, and owns outright one whole end of the world.

The Antarctic belongs to the penguin.

When one sails south from any part of the inhabited world and passes beyond the latitude of the utmost tip of Africa, South America, or New Zealand, he finds himself in the midst of great seas which stretch around the world, broken only by occasional unimportant islands, and everywhere bordered on the south by an endless circle of ice.

The approach to the ice regions of the South Pole are everywhere by the open sea. Thus are they quite different from the approaches to the North Pole region which are almost everywhere by land.

The animals of the world can travel over land into the Arctic and many of them do so. Thus the Arctic has its polar bear, its white fox, and other land-inhabiting beasts of prey. In the summer it is visited by many birds that find, in its brief season of sunshine, a happy condition for rearing their young.

So forbidding is the Antarctic, so cut off by water from the rest of the world, that no land animals live there. So far away, so barren of vegetation, so uninviting is it that few birds even among those that are aquatic visit it.

Once upon a time when conditions in the world were different, the lands of the Antarctic were regions of tropical warmth, inhabited by birds and beasts. As it cooled off a few of these animals survived by adapting themselves to live in the frigid waters round about.

Some beast of prey of that ancient Antarctic, some bear-like animal, became a seal, and today frequents these regions, and is known as the sea leopard. One type of bird, perhaps formerly a half-land, half-water bird like the goose of the barn yard, retained its place in the far South by becoming even more a creature of the water, by copying the methods of the fish.

This was the penguin. Today it is the outstanding element of animal life for this whole end of the world. It had a counterpart in the northern hemisphere in the great auk, now extinct, not a close relative, but one which resembled it in its odd wing structure.

From whatever point around the world one approaches the Antarctic he finds the first evidence of it in the outriders packice which run all the way around the pole at about the sixtieth degree of latitude. Thus the packice of the South Seas is as far from the South Pole as Petrograd, Stockholm, the south point of Greenland, and Seward, Alaska, are from the North Pole. Practically all below this sixtieth degree of latitude is a waste area on the earth's surface, its greatest waste area, and is given over to ice. On its fringes the penguin reigns.

The outriding shore line of this greatest of the world's wastes is a shifting, churning mass of packice, which marks the point where temperature becomes so low that water changes from the liquid to the solid form.

Wherever one touches this region of grinding ice, there are to be found penguins, standing like so many ten pins, the slate gray or black of their backs clear cut against a world of whiteness. They stand there, upright on the ice ledges when the long Antarctic day begins to break, welcoming the sun. Worshipers of that great ball of fire they seem, drawn up here in formal ceremony

A PENGUIN COLONY IN THE ANTARCTIC

against the day of its coming, the only living thing in all this waste to do it reverence.

There are two orders of birds in the world that are ridiculously unlike the great mass of bird life. These two orders are the ostrich and the penguin. Both are birds that do not fly, birds that have but silly little flippers left of what were once good wings.

The ostrich is a dweller in the desert, a bird which got

the habit of running instead of flying. It entirely stopped using its wings, and so they wasted away. Under desert conditions the ostrich became the peculiar animal which we know today.

The penguin, in the face of the advancing ice of the South, according to theory, found that its chance of surviving lay in becoming able to procure food beneath the surface of the icy waters that surrounded it, for, due to increasing cold, food in the air or on the land was ever diminishing. Possibly its wings never were developed for flight, but if they were, it stopped flying and began a more active practice of diving. It certainly has been millions of years since the penguin tried to fly. During all that time it has been using its wings as a help in swimming. Thus its wings are today but a pair of water paddles, and one of the best in all the world. With their aid, the penguin has conquered the Antarctic and reigns supreme beyond the pack-ice barrier.

There are fifteen or twenty different kinds of penguins, but of these the Adélie penguins are by far the most numerous. They are to be found everywhere on the ice rim of the Antarctic, while others appear only here and there. The Emperor penguins are about twice as large as the Adélies, weighing as much as seventy-five pounds, and are strangely different in their habits. There are a few penguins here and there in the South Seas outside of the Antarctic, but these groups are unimportant; most of the tribe are in the far South. The Adélies are the dominant strain, and it is of them that we shall speak.

An Adélie penguin is about two feet and a half tall and may weigh from thirty to forty pounds. It is a squat, upright little figure, unsteadily balanced on short, web-footed legs. It loses little in height when it sits down, propped by its tail, that it may not topple over. Either standing or sitting it is amusingly man-like in appearance. Its head, its flippers, the back half of its body, and its tail are black. The shape of this black part of its body covering suggests a man's evening coat. The front is immaculate white, which, again, suggests a dress shirt front. Altogether, the appearance of this penguin is remarkably like that of a fat, sprightly, little man dressed for a party, and somewhat self-conscious in his evening clothes. Groups of them suggest a party of masqueraders out on Hallowe'en, for there is a grotesqueness about these dress suits that, it would seem, can come only from a desire to be funny.

Its flippers, short bone-like paddles, covered with feathers that look more like scales, are the most remarkable item in the penguin make-up. They were, perhaps, once wings with which it could fly. Wings, the scientists tell us, come from hands. Hands in the beginning, developed from the fins of fishes. If this theory is correct in its application to the penguin, that bird has almost completed the circle from fish to bird and back again to fish.

Still, the penguin swims in no wise as does the fish. When the seals, which were land animals, went back to the ocean, they again took up the fish scheme of swimming. They used their tails like a single oar in the back of a boat and sculled with it.

Penguins swim with their arms as do human beings. They use the overhand stroke, first with one hand, and then with the other. They use their web feet hardly at all in swimming, except as rudders. Yet, in this overhand swimming, they think nothing of starting out on unbroken hundred-mile trips, and race desperately for their lives with their arch enemy, the sea leopard.

The penguins welcome visitors to the South. If those visitors go ashore they are likely to be surprised that these creatures of the wild do not run away, but instead come curiously out to meet them. They have seen so little of men that they have not yet learned to fear them. It is thus with all animals in the regions where man is new. They do not fear him until he has proved himself dangerous to them, a thing, alas. which he is likely very quickly to do.

Often a group of penguins, appearing to be a reception committee, will come out to visit man creatures who enter their solitude. One individual, the chairman of the committee, perhaps, walks ahead of his fellows. He is likely to come very near to you, to move his head back and forth, to peer at you out of first one eye and then the other. Then he is likely to open the conversation, but whether in greeting or by way of question, nobody has ever been able to determine.

"Squawk," he says.

It helps to establish friendly relations if you answer him in kind and respond by saying "squawk." If you do this the chairman seems satisfied. If you do not he is likely to appear irritated and repeat his first remark several times. Soon, however, his interest wanes, and he is as likely as not to yawn in your very face. to settle on his haunches, and go to sleep.

If you visit the Adélie penguin in the nesting colony, however, your reception is quite different. There, also, there is no fear of you, but you meet with unfriendliness everywhere you go. From every nest an angry bill reaches out to nip you about the ankles. Those birds that are not on the nest may meet you face to face and try to block your way. The pinching of bills and the pounding with wings is sufficiently severe to leave bruises that will be quite blue on the morrow, but beyond this you are uninjured.

July is midwinter in penguin land. During that month and for three or four others on each side of it, the penguins are out here on the fringe of their waste continent, foraging a livelihood from the icy waters or keeping guard on the outlying ice ledges. They frolic like porpoises in the water round about, making a play ground of the open sea for half a hundred miles. Here they find abundant food in the form of small fishes and crustaceans, such as shrimp. Here they play hide and seek with their two desperate enemies, the sea leopard and the killer whale. This sea leopard, a seal eight or ten feet long, preys upon the penguins in the water, just as its ancestors on land may once have preyed upon their land-bird forebears. An examination of the stomach of one sea leopard may reveal remnants of as many as ten penguins which it has eaten.

The killer whale, twenty feet long, one of the most ferocious animals in the world, so vicious that it readily devours the sea leopard, attacks that largest living thing, the whalebone whale, and tears it to pieces, frequents these waters. It prizes highly this penguin which makes a neat thirty-pound morsel, greasy and nourishing, like a sausage, on its bill of fare.

If one would know the penguins, however, he must visit them on a midsummer day in January, when they have retreated far south to their nesting place and are in the midst of rearing their families. As summer comes on, penguins go on a pilgrimage far to the South, where the ice is breaking up and certain rocky islands are emerging for a brief time from their blanket of snow. These islands are the nesting places of the penguins. It is here that they are at home. It is here that, now and again, in the past century, scientific expeditions have come, have studied this strange penguin civilization, and have reported it in detail.

In midsummer this island home of the penguin is surrounded by open water, but at the time when the advance guards of penguins arrive, it is still snow swept and there may be ice fields about it for scores of miles.

The penguins swim wherever they may on this trip home, for they are clumsy travelers by land. When they must walk, they waddle awkwardly on their duck-like legs and their progress is tediously slow. Despite this fact, however, they often cross scores of miles of snow and ice that they may arrive early at the nesting ground.

Few sights in the world are more impressive than the progress of one of the Adélie penguin caravans through the waste snow stretches of the Antarctic. They appear in man-like columns, walking single file and waddling doggedly ahead. Then, to the surprise of the onlookers, they adopt an entirely new and strikingly novel method of travel. On a down grade or where the snow is right for it, they drop upon their breasts and toboggan or kick themselves along with their feet. They become animated sleds instead of marchers. Their glossy breasts seem to make good sled runners. Where the snow is smooth they seem to get along better as sleds than as marchers. In this manner of going they use their wings in steering, and otherwise helping themselves along, although most of this sort of progress is made by pushing with their feet. After sledding it for a while they again stand upright and walk. In this way, by changing from one method to the other, they seem able restfully to break the monotony of their journey.

At the nesting grounds the female Adélie penguins take the first steps toward establishing a home. Each selects the place for her home in this favored area where the winds blow so fiercely that no snow has gathered. The ground is still frozen, for the thermometer registers around zero, so the female penguin sits on her home site, that the warmth of her body may thaw it out. As the surface thaws she scoops it out, until an oval basin, two feet across, has been made, as the beginning of what is to be her nest. Here she awaits the coming of her mate.

The males arrive overland very much exhausted by their long and tedious journey. As they approach the colony the different arriving birds give every evidence of joy. They break their lines, scramble wildly, wave their wings about, and scream noisily.

By following the actions of one of these newly arrived males, the observer may trace the penguin courtship through its course. This male, evidently very tired, looks over the nesting ground, then in preparation for his campaign, props himself up on his legs and tail, relaxes, and takes a brief nap. This lasts but ten minutes or so, when he awakens suddenly, pulls himself together, and starts very energetically about the business in hand.

He approaches the colony where the waiting females have scooped out their nests. He may walk about one and another of these females, which, true to sex instinct, appear unconcerned, and inspect them closely. It is hard for the man observer to know why this newcomer decides not to accept one and another of these females. He is likely, however, to go on and on until he has seen four or five before he makes a choice. Then at the approach of some penguin lady, apparently remarkably like her sisters, his mind is suddenly made up. He seems to know at first sight that this is to be his mate.

In penguin land, pebbles are legal tender. It is pebbles that are more precious than anything else. It is pebbles of which the nests are made and the supply is usually less than the demand. Unless there is a pile of pebbles as a foundation for the nest, the melting snow of midsummer will run into it and spoil the eggs. The pebbles support the eggs above this water.

When this male bird, therefore, has decided upon his mate, he hurries away, gets pebbles, and lays them before her. She maintains a dignified reserve, paying no attention to the advances of this stranger. One by one he brings these pebbles to her and presents them as evidence of his affection. By the time that he has brought the fifth or sixth pebble, this lady of his choice is likely to begin to relent, to take notice of her wooer. Responding to this, he stretches himself proudly before her as if for inspection.

She seems to be duly impressed and they put their heads together, squawking noisily.

The two now proceed to build their nest, the male bringing the stones, and the female, first carefully paving her dug-out, and then piling upon this foundation several inches of these loose pebbles. Here she lays her two eggs and here she and her mate take turns at sitting on them until they are hatched.

ADÉLIE PENGUINS QUARRELING

Soon the entire wind-swept waste which is the nesting ground, is thickly covered, and may have a population of as many as one hundred thousand penguins. The nests are not more than two or three feet apart.

In the weeks that follow frequent recurrences of two misdemeanors enliven the penguin colony. In the first place there are many males that are late arrivals and they go boldly about, trying to drive established members of their colony from their homes, and take them for themselves. These would-be home-wreckers get into many fights. The intruders are in no way lacking in boldness and in disregard for the rights of their fellows. No penguin, however, fails to defend his home, and so there is constant struggle in which beak and wing furnish the weapons, and in which there is much bruising, no little bloodshed, but very seldom loss of life. Even in a penguin colony there seems to be strength in justice, for the intruder is nearly always defeated.

Always, however, there is a certain roguery going on in the nesting grounds. Theft is the second current misdemeanor in penguin land. These seemingly respectable heads of families, it must be admitted, are given to stealing. There is one article, which, according to their view, it is a virtue to steal. That one thing is the precious pebbles with which they build their nests.

One who watches the well-formed penguin colony on a quiet day will see now and then some individual slinking among the nests, his feathers hugged so close to him that he seems much smaller than usual. Watch such a bird and you will presently see him approach stealthily from the rear the nest upon which some other penguin is sitting. Most cautiously he tries to steal a pebble from it. If the sitting bird is watchful, she clamors loudly, pecks at the intruder, and drives him away. This thief does not give battle, but slinks away quietly. He seems to know that he is in the wrong, and does not have the heart to fight for so miserable a cause as that upon which he is engaged. If, however, he finds a careless nest holder, one who is not watchful, he steals a stone and bears it away to his home. Encouraged by this success, he returns and gets another and another. The nest holder who is not watchful is likely soon to find that most of her house has disappeared. There is likely not to be enough of it left to keep her eggs out of the water when the thaw comes. Her eggs will spoil. Thus do those lacking vigilance fail to produce young.

The penguins eat nothing during this first month at home-making, so become gaunt, dirty, bedraggled. Then, after the eggs are laid, they begin to straggle away to the water edge. Always a sociable lot, they gather in groups and trudge away to the ice foot. They gather where the ice overhangs the water, like boys at the swimming hole, but hesitate to plunge in. Those behind begin shoving, try to push those in front overboard. These wriggle out, run around to the rear, and play the pushing game. Finally, after an hour of scuffling, one plunges in and the others follow in rapid succession. They frolic in the water, a favorite game being follow-the-leader. This leader goes careening about, cuts fancy figures, dives beneath an ice ledge there, leaps upon one here. These leaps are remarkable. A penguin may leap from the water ten feet out from an ice ledge that is six feet above the water and land securely upon it, the procession following him. Then they are off again in endless play. A piece of floating ice may come by in the current. The penguin herd will swarm over it, stealing a ride as might so many tramps on a freight train. A huge iceberg may strand on the beach, and unceasing strings of these penguins, for days at a time, will engage in a man-like manner in the sport of mountain climbing.

Finally the young are hatched, and the job of the colony becomes that of keeping them in food, for they have immense appetites. The mother and father take turns in going out to sea, gorging themselves with shrimp and waddling home. Many birds in this world feed their young by bringing up food that they have borne home in their crops. Most of these put their bills inside the bills of the little ones, turn the stop and thus fill them up. The penguins, perhaps to be different, take the bills of the young inside theirs. This method does not seem to be so good as the other. There is more leakage.

Finally, the young penguins are big enough to leave the nests. They then become members of a rabble that inhabits the waterfront, and clamors for food. The mothers no longer know their own children. Any old bird feeds any young one that presents itself. It is the task of the old birds of the colony to fill the young ones and they work at it and seem to succeed, for most of the new generation live through their childhood.

Now the summer is drawing to a close. February has come and the midnight sun is retreating. The old birds go into seclusion for a while, shed their feathers, get a new coat. All is in readiness for leaving the nesting grounds. Gradually the old birds and the maturing young ones strike out to sea with their overhand swimming, roll away to the north for hundreds of miles, take their places there on the outer rim of the packice, sentinels to welcome whatever mariner may chance to drift this far, away from the customary haunts of men.

QUESTIONS

1. Describe the peculiar location of the home of the penguin. How would you get there? Why are there fewer animals in the Antarctic than in the Arctic?
2. Ages ago Antarctic regions were warm. What became of the animals as this part of the world grew cold? What were the two most striking types that survived? How did they manage it? Describe the sentinel penguins on the packice.
3. How are the ostrich and the penguins alike and unlike? How and why did the penguin change from a flyer to a diver? How did its wings change?
4. The Adélie penguins are the most numerous. Describe them as to size and appearance. How do they swim?
5. How do penguins regard human beings? Why are they not afraid of man? Describe their "reception committee." How is the reception different in the nesting colony?
6. July is midwinter in penguin land. Where and how do they live at that season? What are their worst enemies?
7. Where do they go for the summer which in the far South is January? How do they reach the nesting grounds? Describe an Adélie caravan traveling overland.
8. How are the nesting places selected? Tell of a penguin wooing. Of the strange use of pebbles.
9. Describe the game of pebble stealing. Why are pebbles so highly prized?
10. When do the penguins act like boys at the swimming hole? What do they do? Describe some of the games they play.
11. How are the young fed? On what? Describe the clamor of the colony as the young ones grow up. When does it break up?
12. There are many strange kinds of birds all of which seem to tie back to that common ancestor that learned to use wings for flight. What birds have you seen that seem to you as strange as the penguin?

CHAPTER IX

THE AMERICAN ROBIN

THE robins sing in the rain.

That fact gives a clue to the sort of birds they are—hardy, bold, fearless, irrepressible. You cannot keep a robin down. He is always fat and cheerful, saucy and daring, jolly and sociable, busy and buoyant. He is the perfect example of the sort of person anyone would like to make his friend and keep near him, of the genial and bubbling spirit that makes the world a pleasanter place in which to live.

The robin calls lustily in the garden. "Be brave!" he says. "Be brave!" Then, after sunset, he twitters with a milder theme in mind and what he seems to say is: "Be more gentle; be more loving." His salute to the sunrise is softly warbled, and his evening song is pensive. In his courting days his voice is full of rapture and joy, while the song to his brooding mate is low and sweet. He can express joy, hate, fear, alarm, happiness in that voice. Yet always it dominates. It rises above the chorus of other bird voices as the notes of the soprano soloist swell out above the others in the great church choir.

The robin is an outstanding figure in the bird world, about the most wholesome, hearty, robust, sociable creature of them all.

Human beings generally prize the good qualities of the cheery robin, and to show their appreciation, treat him well. There was the case of the engineer, for instance, who was building a power plant. He had set his steam shovels to work eating a big lot down to level as a basement for his building. Then he found that a robin had built its nest in a bush at the center of the lot. The young birds were just hatched.

The engineer ordered his shovel-men to dig out all around this bush, but not to disturb it. It remained there on its quiet island in the midst of much clamor. Finally, with the young birds just coming into their pin feathers, a time arrived when the job must stop or something be done with the robins. This engineer then had a platform built about the bush that bore the nest, had it securely braced and its trunk sawed off. While the great steam shovels waited he thus moved the whole, uninjured, to a zone of safety.

Only the boy with the sling shot shows no mercy to the robin. A mother robin, for instance, had built her nest in the back yard of a town house and was well thought of by her host. The boy next door, however, could not resist taking a shot at her, with the result that she fluttered helplessly with a broken wing. The owner of the yard she had chosen bound a splint on the wing and put the bird in a cage. He wondered what would become of the eggs.

Father robin rose to the emergency. He took his mate's place on the eggs, and brooded them very carefully. It had been his habit to bring worms to his mate that she need never have to leave the nest, but now he must steal away and hustle his own food, being careful to get back before the eggs were chilled. Even with this double task in hand, however, he insisted on feeding his wounded mate. As she swung there in her cage he brought her many long and juicy worms, and handed them through the bars.

YOUNG ROBIN LEAVING NEST

He hatched the young ones and they were half-grown before the mother robin was able again to help.

The American robin, strange to say, is a quite different bird from the robin redbreast of Europe, a feathered songster of which much has been written and which has a place in the affections of people of the Old World that is very similar to that of its namesake in the United States. This is the bird that covered the "Babes in the Woods" with leaves while they slept, and that, on another occasion, was killed by the sparrow with his bow and arrow.

The robin redbreast is the original robin. It is the bird that goes back to the word "rubeo" meaning "red" for its name. It is the bird of which the legend comes down to the effect that it was present at the crucifixion of Christ. In pity, it flew to the suffering Saviour, there on the cross, and sought to relieve him in his suffering, trying to pull one of the thorns from his forehead, thorns from the mock crown that had been forced down on his brow. It was then, goes the legend, that a drop of the blood of Christ fell on the breast of the bird and made it forever crimson. It was then that Jesus blessed the bird and made it take a message of joy to generations yet to come.

This European robin is a much smaller bird than the American; is, in fact, not so big as an English sparrow. It is but six inches long, while the American robin, which is so well known that it is taken as a standard by which to measure other birds, is about ten inches long. The European robin, however, is much more brilliantly colored than the big fellow of the west.

The American robin is a thrush, while the European bird is not a thrush at all. The American is a cousin, on one side, to that sweet and solitary singer, the hermit thrush of the dense woods, and on the other to that symbol of happiness, the bluebird. Yet it has a chestnut waistcoat, suggestive of the redbreast, and the same sort of cheerful and friendly manners. So, when English settlers came to America and saw this bird, they gave it the name of their old friend, though the birds are entirely different. Thus Englishmen wonder at American descriptions of robins, and Americans at those of the English. Science, too, is confused, for "robin" means one thing on one side of the world and another thing on the other.

The robins of America range from the Gulf of Mexico to Labrador, blanketing the country with their presence. They are but slightly migratory. They are to be found both summer and winter in the southern half of the United States. To the northern half they come in the very early spring. They follow the disappearing snow all the way to the Arctic, and are almost the first birds to appear as heralds of spring. The fact that they come so early adds much to their popularity. The world that is snowed in is brought word by the robins that it is soon to be released.

The male robins come first, in flocks. The size of the flocks grow less as they pass to the north, individual birds dropping out along the way at their old nesting places. Then, a week or two later, the females arrive and the business of home building begins.

The nests are likely to be made in low-branching trees, not more than ten or fifteen feet from the ground, in fences, in shrubbery about the gardens, in the vines that grow by the window, even under the shelter of the front balcony. The nest is a rough-built sort of structure, beginning with coarse grass, roots, and leaves, lined with mud from some near-by stream, and finished with soft materials. It is this mud element that makes it necessary that the robin's nest should be as much protected as possible from the rain, otherwise it may melt. It is this use of mud, also, that causes the mother robin to come to have such an unkempt appearance during the nest-building period. She smooths the mud lining with her breast and so becomes quite soiled. This labor past, however, she goes for a bath, a thing of which robins are very fond, plumes herself carefully, and is again neat and prim.

The robin lays four or five eggs in its nest, thereby indicating that it is a bird that faces about the average number of difficulties in rearing its brood. That is about the number of eggs that is laid by the gull, for instance, which faces average dangers, and by the plover, the warbler, and such woods birds, by the screech owl, and by many others. It is a strange fact in nature that the more danger of death birds face, the larger the number of eggs they will lay. If there are many dangers there is the need of bringing more young birds into the world so that enough of them will survive to carry on the race. If there were insurance companies in the bird world, they might start out with a fair measure of the danger of death to any bird by counting the eggs in its nest. The more eggs it lays, the greater is the danger to the individual bird. In nature the birds seem to sense the danger and lay the number of eggs necessary to keep the numbers up.

Many birds do not find it necessary to lay so many. That is because circumstances are such that not so many eggs or birds of their breeds come to grief. The mourning dove, for instance, lays but two eggs. That is all that is necessary because the dove is so swift of flight and the seeds on which it feeds are so plentiful that its chances are good for keeping alive. The nighthawk lays but two because its eggs are so nearly the color of the ground on which they are laid, and the bird hides itself so well by the mottled coloring it adopts, that it does not suffer great danger. The humming bird lays but two eggs because it is so quick on the wing as to require little life insurance. The eagle, being the master, needs to lay but two or three eggs to keep the race going. Many murres, auks, gannets, and other sea birds lay but one egg. That is all that is necessary because they live on lonesome rock ledges where there is little danger.

On the other hand, there are many birds that find it necessary to lay greater numbers of eggs than the average. This is particularly true of those game birds like partridges, grouse, prairie chickens, pheasants, ducks, and geese, which are attractive to beasts of prey and which are comparatively easy for them to catch. These birds find it necessary to lay from ten to twenty eggs to keep their families going. Such big birds as the ostrich similarly lay great numbers of eggs. It is an odd fact that these big desert birds, ostriches, rheas, and emus, living on such continents as Africa or South America, lay more eggs than they do when living on isolated islands in the South Seas. This is because, on the continents, there are more beasts that prey upon them than on the islands.

The infinite variety of birds' eggs has been the thing that has tempted boys through the generations, thoughtless of the destruction to bird life that resulted, to make collections of them. Much greater would have been the interest of those boys had they known that there were special reasons back of the special colors of all those eggs. The reason was nearly always that nature made them resemble their backgrounds so that they might not be seen by those creatures that would eat them if they could find them.

There are certain eggs that are pure white, such as those of the wood-peckers, which dig their nest out of tree trunks, those of the burrowing owls that nest under ground, and those of the king-fisher, which tunnels into a bank. Were they out in the open they would be found by the first egg-eating snake, crow, or squirrel that passed their way. Being placed in holes, however, they need no color protection. To be sure there are birds, as, for instance, the doves, that lay white eggs out where all may see, but these cases are exceptional.

The cassowaries, huge birds of New Guinea, lay green eggs, six inches long, in the jungles where they live, matching the grass so exactly that they are very hard to find. The olive gray egg of the tern, with dark blotches, lying on the beach among pebbles and seaweed, can hardly be seen. The black skimmer of the Atlantic Coast lays a similar egg among the old clamshells where it is well concealed. The eggs of the hermit thrush, hidden in the forest, are greenish blue. Those of the mocking bird are blue splotched with brown. Those of this American robin, also, are greenish blue.

There are certain odd facts about the shapes of eggs. In the first place there is great strength in their oval shape. A strong man may take the egg of a hen between the palms of his two hands and may exert all his strength on it, and yet not crush its seemingly frail shell.

The fact that the egg is bigger at one end than at the other, also, is very important. Some eggs are, in fact, much more top-shaped than others. Take one of these sharp-pointed eggs, put it in the middle of a board a foot wide, and try to roll it off. The egg when rolled describes a small circle but remains on the board. Most nests are high from the ground and Humpty-Dumpty must not have a fall. This is particularly true of such cliff-dwelling birds as murres, which lay their eggs on rock ledges. Their eggs are so shaped that they roll least of them all.

Withal, birds have developed the egg as a means of producing their young to a higher degree of perfection than any other of Nature's creatures. The bird got much egg education of course, from its ancestors, the reptiles. Reptiles have highly developed, round eggs with a tough, non-brittle shell. They lay few eggs and take good care of them. Farther back, down the animal scale, are the fishes, quite wasteful users of this idea of eggs as a cradle for their young. The codfish to this day lays millions of eggs that it may produce one new fish like itself. Its egg is one of the worst insurance risks in the world. The mourning dove, even the robin, has done much since the time when their ancestors came out of the water to increase the safety of eggs.

It is while there are young ones in the nest that the cock robin is at his best. It is said by those who are supposed to know, that a young robin eats two or three times its weight in worms and insects in a single day. It requires about four feet of earthworms to ration a young robin for one day. For four young robins the parents must hustle sixteen feet of worms.

Cutworms, caterpillars, moths, grasshoppers, and beetles make up a considerable portion of the young robin's food, but earthworms are its special delight. The elder robins are very wise as to the habits of this pink bit of animated bird macaroni and its behavior beneath the surface of the lawn. Earthworms come out to look around after a rain, but, unfortunately for the robins, it

ROBIN STEALING A CHERRY

does not rain every day. When the owners of close-cropped lawns sprinkle them with the hose, it is almost as good, and robins know this. Even without sprinkling, the lawns offer quite good hunting. Cock robin goes hopping about in a jerky way quite peculiar to himself, pausing now and then to look about for his prime enemy, the treacherous house cat. Then, suddenly, he may be seen to stick his neck straight out, to listen intently, to cock his head slightly on one side. Finally he strikes, and comes up with an earthworm in his bill. These earthworms, having no eyes, anchor themselves to their holes by swelling out the rings at the end of their bodies so they hold tight to their doorways. Because of this the robin may have to do a bit of pulling and tugging before he gets his prey and he may appear rather ridiculous while he is doing it. He seems, however, to like this game of tug-of-war. Soon the worm's hold breaks and the robin is away with an item of breakfast to the young ones in the nest.

Despite the fact that the robin is so dearly loved, there has long been local resentment against him because he sometimes does injury to certain fruit. He has a very strong liking for cherries. Strawberries also appeal to his palate. When the cherry crop or the strawberry crop is limited, and wild berries run short, the robin is likely to become destructive. He may eat up all the cherries. As long as wild berries are to be had, however, he does not injure man's fruit. He likes the wild fruit best. Sometimes clearing up the waste land does away with the wild berries and the robin is driven into small orchards and takes all the fruit. This is a hardship on the individual. From the standpoint of the whole community, however, the robin is beneficial, for he has a huge appetite and it is largely satisfied upon those insects that are injurious to crops.

The robin does not stop at one brood of young ones a year. There are usually two and sometimes three. By midsummer the mother robin is preparing for her second brood. The father, however, is busy with another duty. To him has fallen the guidance of the older children, those youngsters whose speckled breasts carry to the world the message that they are thrushes. The first brood follows him about the fields, hunts and frolics all day long. When nightfall comes they do not return to the vicinity of the nest, but troop away to a general gathering point which has been established. This is usually in some growth of young trees along a stream or in some marshy region. All the young robins for miles around are led here by their fathers. The colony grows steadily. The father robins go back to their mates by day and help in feeding the new family, but by night they return to this community center for their kind.

Finally, when the second, and possibly the third, brood is added to the flock, the time is near for the migration. The robins are becoming shy now, are taking to the tree tops. The molting period comes on and they hide in the woods while getting a new suit of feathers. By this time it is autumn and the trees are red like their own breasts. The gusts of winter scatter the leaves and with them, in the northern states, go these birds. They fly away for the winter, to a latitude that is below the 40th parallel south of Baltimore and Kansas City. There they are likely to congregate in cedar groves, for in them there is both shelter and food. There are dense populations of robins, for instance, in the winter time in the cedar brakes of Tennessee.

It used to be that hunters killed tens of thousands of these wintering robins and sold them in the markets of the cities. This and the boy with the .22 rifle, and hunters everywhere, some years ago hammered the numbers of robins down until it looked as though there would be none of them left. Then a law was passed putting them among the protected migratory birds and now it is coming to pass that robins are getting more numerous, that they are becoming common among the old apple trees throughout the North that of old knew them so well.

QUESTIONS

1. What sort of bird character is the robin? What kind of song does he sing? What does he say when he sings?
2. Tell the story of the engineer who saved the robin's nest. Of the boy with the sling shot. Of the robin's care of his wounded mate.
3. How does the English robin redbreast differ from the American robin? Which is the original robin? What is the legend of the origin of the red breast of this bird?
4. To what well-known group of singing birds does the American robin belong? Name some of its cousins. Is the English robin a thrush?
5. To what extent are the American robins migratory? How do they happen to appear first in the spring? Describe their coming.
6. Where and of what do they make their nests? How many eggs are laid? What fact of a bird's life is shown by the number of eggs it lays? How do you explain this?
7. What are some of the birds that lay fewer eggs than the robin? What birds lay more? Why? Why do birds of the ostrich type lay more eggs in Africa than they do in the South Sea islands?
8. What is the color of the eggs of birds that nest in holes? Why? What is the reason for the various colors and spots of eggs? Give some striking examples, from your own observation if possible.
9. Why are eggs shaped as they are? From what ancestors did the birds get the egg-laying habit? Show the wastefulness of eggs of the fishes. The economy of the birds.
10. What can you say of the appetite of the young robin? Describe the robin's hunt for earthworms.
11. What part of the robin's bill of fare causes the farmer to dislike him? What drives him into the cherry trees?

12. How many broods of young robins are hatched in a season? What training do the young robins of the early brood get from their father? How is the robin colony built up in the late summer?
13. Why do the robins hide in the woods in the autumn? Where do they go in winter? What do they eat?
14. Market hunters used to shoot great numbers of robins. How was this stopped? What can be said of their present numbers?

CHAPTER X

THE CHICKEN

THE most productive bird in the world is the barn-yard chicken.

The chicken, in fact, produces more wealth, year by year, than do all the other wearers of feathers combined.

For the matter of that, the chicken produces more wealth in the world than does any other living thing, animal or plant, except possibly the cow, and man himself.

It produces more wealth in the world than do gold mines, silver mines, diamond mines combined.

The wealth produced by the chicken in the United States alone amounts to more than a billion dollars a year. This is greater than the yield of all the wheat fields, is about equal to the yield of the cotton fields, is half as great as that of the corn fields. In all the countries of the world together, the chicken far surpasses cotton or corn as a producer of wealth. In the United States, cattle, considering both meat and dairy products, yield nearly three times as much wealth every year as do the chickens. Considering the whole world, however, it is doubtful if Bossy has the lead, for there are huge areas like interior China where chickens are plentiful and the cow is almost unknown.

So familiar is the chicken that few people ever stop to consider it alone, to stand it up for a look, to measure its importance in the world. It does not get its due. It is even, at times, an object of ridicule. And all the time, widely scattered as it is, producing its steady flow of eggs for breakfast, broilers, and grown birds for roasting, it is laying firm hold on the claim to being man's best friend.

Neither do people stop to consider the origin of this familiar bird, whence it came, and by what routes. Nobody asks to whom the world should give thanks for this rare gift.

To begin with it may be set down that the chicken was given to the world by none other than distant India. Every chicken in every land traces back to India just as every blue-eyed human being ties back to North Europe and every slant-eyed one to the east of Asia.

Today the jungle cock in the wild state, looking for all the world like a brown leghorn rooster, crows along the road to Mandalay. Anywhere in that forest waste that extends from the Himalaya Mountains down through eastern India, through Burma, Siam, and the Malay Peninsula, the traveler may hear the familiar cackling of the little brown hen.

Here the chicken is wild and in its native haunts, living much in the same way as do our quail. This jungle fowl, though it knows it not, there in its wilderness, is the granddaddy of all the barn-yard, wealth-producing chickens in the world.

It was on the borders of these forests that the yellow men whose eyes are slant, and the black men whose hair is straight, emerging into an eastern civilization when the western races were still savage, caught the wild chanticleer of the jungle, tamed it, and taught it to make its home about their dwellings.

The first written record of the chicken comes from China. Here fourteen centuries before Christ, while the ancient Greeks were still wild men, the Chinese wrote that they had secured this fowl from the west. This meant that it had come from India. More than a thousand years later, after the Greeks had learned to write, after the time of Homer who never mentioned chickens, they wrote of the introduction of the Persian bird. It seems that the chicken had come to Europe from India by way of Persia. From then on it found its way wherever men traded with one another and today has a firm place in the homes of all the stocky peasants that till the fields of France, of all the coolie farmers that make up the uncounted millions of China, of all that host of settlers that have converted the wastes of America into fertile farms.

But in the forests of India the jungle cock still announces the coming of day with his cock-a-doodle-doo, calls his women folk to him when he finds a worm, struts about in his unbounded conceit, and often fights to the death with the rival cock that disputes his leadership, just as do his grandsons all the world around.

We know that the domestic chicken and the jungle fowl are one and the same because, in the first place, they have upon their heads that red flesh comb, a mark that is peculiar to them, that is carried by no other bird in the world. Both chickens and jungle fowls have, further, two wattles hanging beneath their chins such as no other birds possess. They are both armed with death-dealing spurs on their legs, a fighting equipment that is carried only by them and a few of their cousins, the pheasants, the peacocks, and the turkeys. Chickens and these wild birds are built alike in every way. Both crow and cackle and otherwise disport themselves like blood brothers. Finally, when the domestic chicken and the jungle fowl are brought together, they mate with one another and bring forth young ones just as though they were members of the same brood.

These birds, according to the scientists, are of the "galliform order," which means that they are hen-like, or fowl-like birds. It easily can be seen that they are quite different in form and habits from the birds of prey, for instance, from the woodpeckers. They live mostly on the ground, are good at running, poor at flying, and eat both grain and insects. The quails, grouse, prairie hens, pheasants, turkeys, and guinea hens are all members of this order. None of these latter, however, have the sort of comb and wattles that distinguish the jungle fowls and the barn-yard chickens. None of them will mate with them.

Most of the domesticated birds are members of the "galliform order." The peacock, which also comes from India and which was introduced into western Asia by Solomon, is a member of that order. So is the guinea hen which comes from Africa. Ducks and geese are of another order, but the prime table bird of them all, the turkey, belongs here.

It is an odd thing that there are but sixty of the members of the animal kingdom that have been domesticated since time began. The other hosts of the wild do not fit into man's scheme or refuse to become a part of it. Nearly all of those that came under the sway of man did so when the race was young. There are almost no animals that have been tamed since written history began. Almost the only exceptions are those creatures found in America after its discovery that lent themselves to domestication, but that had been little under the influence of man before the coming of the Europeans. Of these there were three, the llama, beast of burden of South America, the guinea pig, and the turkey. This turkey, found widely distributed through North America, has proved itself the choicest of all table birds and has found a place pretty well around the world. As India contributed chickens to the world, so America contributed turkeys.

This order of the hen-like birds probably supply the best example there is of the peculiar arrangement in the feathered world for the digestion of food.

Most mammals have teeth and grind their food more or less before sending it on to the stomach for digestion. This process is varied in a manner that approaches that of the birds, by the cud-chewing animals such as the cow. These animals go out into the pasture lands where the grass is abundant, rapidly eat great quantities of it, and store it away in a preliminary stomach known as a paunch. Then they return to the shade, lie down comfortably, and their secondary stomachs pinch off quantities of the grass mass that they have stored away, much as mother used to pinch off bits of dough for making biscuits. They belch it up, chew it thoroughly, and swallow it again.

Most of the birds also have a bag in which to store their hastily eaten food. This is called a crop. It is, in fact, but an enlarged chamber in the lower part of the gullet, and is suspended quite outside of the bird's body chamber, hanging to it as does a balcony to the front of a house.

This crop is developed to an unusual size in most of the grain-eating birds. The birds do not have teeth with which to grind up this hard food, and the crop, aside from serving the purpose of storage, moistens and softens it.

After this preliminary preparation the food is passed on first to a lesser stomach where it is mixed with digestive juices, and then to the gizzard, which is a modification of the stomach, peculiar to the bird and a really remarkable structure.

The bird chews its food in the gizzard. As a substitute for teeth it fills that gizzard with small stones or gravel which serve for purposes of grinding. The gizzard, which is a surprisingly sturdy body of muscles, by its contractions grinds this food and gravel together in a way that gets remarkable results. It grinds up corn, wheat, acorns, beetles, bones.

There is an odd fact of nature introduced just here. Some of the reptiles also have gizzards. The alligator, for instance, has one in which it may carry grinding stones as big as one's fist. There is much evidence that birds developed from the reptiles, and the fact that both have gizzards is but another bit of this evidence.

Experimenters have mixed up jagged pieces of broken glass and forced fowls to swallow them and have found that the gizzard grinds them down to harmlessness. One experimenter arranged a lead ball with needle points sticking out of it all round and induced a turkey to swallow it. He later killed the turkey to see what had happened to these needle points, and found that the gizzard had quite completely ground them off. He repeated the experiment with the points of knife blades sticking from the lead ball, with the same result. In none of these experiments had the tough leathery lining of the bird's gizzard been injured.

The jungle fowl of India has undergone many changes, many modifications, because of the influence of man. In the native state, for instance, they are all of a color—black breasted and red backed. The varieties of game chickens most ordinarily met are of about the same color as the jungle fowl. Man has so bred these chickens, however, that he has developed an almost endless variety of color and form.

Nowhere has man more frequently demonstrated the possibilities that lie in selective breeding. One may, for instance, have a flock of Black Spanish chickens, a well-established breed. In the flock there may be two chickens, a male and a female, each having a single white feather in its coat. If these two are crossed, some of the chickens are likely to have three or four white feathers. If those with most white feathers are crossed, there may be in the next generation chickens with still more white feathers. Continuing to select and cross the chickens with the most white feathers, the experimenter will eventually produce from this black stock chickens that are all white.

By selecting to any given color, a race of that color may be produced. It is thus that chickens of so many colors have come to pass, all from a stock which, in the wild state, was colored like the brown leghorn or the game chicken of today.

By this selective breeding a very small race of chickens may be created, as, for example, the bantams. Again, by selecting for large size a race of giants may be created, as the Brahmas. By selecting for details of build and disposition a fighting cock may be developed, as the game chicken. By selecting those that tend toward crests a race may be developed with huge clumps of feathers on their heads as in the Polish breed.

In view of these known possibilities of selective breed-

GAME FOWLS FIGHTING

ing, it is interesting to note just what has been developed from this little jungle fowl, but slightly bigger than a partridge.

There have been two lines of most striking and curious development in chickens, one extending east from India and unfolding itself chiefly in China, and one extending to the west and working itself out principally around the Mediterranean. Today the two outstanding races of chickens in the world are the Asiatics and the Mediterraneans.

Those chickens that spread westward into Europe have been the constant companions of these western men for two thousand years. Through most of that time man has not consciously applied a great amount of selective breeding to them. In the west he has but recently learned the secrets. His first efforts along this line were given to the development of fighting chickens, for matching these has been a favorite sport in Mediterranean countries since the days of early Greek and Roman greatness. Game chickens early became a race apart.

The chicken, throughout Europe, long ago became one of the main dependencies of peasant households and was everywhere to be found. This European bird shows, throughout, a general similarity of type. It is small in size, compared with the Asiatics, and is active and nervous. It is able to hustle a living for itself, and on the farm costs little to keep. It lays many eggs and is little inclined to set. As an egg producer it is the outstanding stock of the world.

In this connection there is another fact that shows the possibilities of developing certain traits by breeding. The jungle fowl of India lays but one nest of eggs a year. There are from seven to twelve eggs in that nest. The Leghorn hen, however, bred for laying, may work at it almost constantly, and may produce from two hundred to three hundred eggs in a year.

While Europe was developing this type of chicken, China was following an entirely different line. Their selection was for size, for inactivity. The chickens of

China came to be twice the size of those of Europe, great hulking fellows that were later to be introduced into the west and to be known as Brahmas, Cochins, and Shanghais.

In the end, however, the master development of the chicken was to take place in America. The old-fashioned chickens in America a generation ago were largely of the

RING NECKED PHEASANTS RISING

Dorking strain, for many centuries a favorite English breed—short-legged birds with white backs and black breasts, with the odd feature of a fifth, quite useless toe, at the back of the foot. Side by side with the Dorking a hundred years ago was to be found the Dominique, that mottled chicken which, crossed with an eastern breed, the Java, produced the Plymouth Rock, long a popular utility breed.

Ships that plied the many seas, together with an awakening knowledge of the possibilities that lay in selective breeding, led to the making of chicken history in America. Nearly a hundred years ago, for instance, Italian sailors brought to American ports a typical, active breed of chicken which was given the name of Leghorn, because it was believed to have come from the port of that name near Genoa, where Columbus was born. The Leghorns proved to be great layers. In America the best laying strains were carefully selected and bred in such a way that the number of eggs in a year was steadily increased. In the end a breed was established that produces more eggs than any other breed of chickens in the world. The Leghorns have become the prize egg-laying chickens of the world. They have been taken from America back to Italy and are displacing the Mediterranean strain from which they came. They are finding their way into all the barn-yards of the world where an output of eggs is the object, and are establishing their mastery.

It was not until after the ports of China were thrown open to the world in 1843 that a knowledge of the giant chickens of that country began to get abroad. These chickens were brought from China by an English traveler in 1853 and presented to the Queen. They became the foundation in England of that famous Brahma strain of chickens. Cochin chickens had already been similarly introduced.

Both Brahmas and Cochins were huge chickens with masses of feathers coming all the way to the ground and hiding their feet. As their numbers increased, they became very popular in England. Later, groups of them reached American ports, usually on vessels from the Orient and somewhat by accident, and formed the basis of those great meat-producing breeds that have largely replaced the former stock.

In America, however, the desire has been to work out combinations of the Asiatics and the Mediterraneans that would keep the good qualities of both and get rid of the faults. The egg-laying Mediterraneans, for instance, were inferior table birds, while the meaty Asiatics were too sluggish, too clumsy, too much inclined to set and neglect egg production.

The Plymouth Rock was the first important result of the mating of the East and the West. It had meaty drumsticks and was at the same time a fair layer and active enough to hustle for its food.

A later development were the Rhode Island Reds, which grew into great popularity. The Buff Cochins, Asiatics, were probably the basic stock for this breed, but they are believed to have been crossed with Wyandottes and Leghorns. Thus was a blend secured that combined egg-laying, meat-producing, and hustling qualities in about the desired proportions. A standard cock was produced that weighs about eight and a half pounds and a hen that weighs six and a half pounds.

Now comes the Black Jersey Giant, admitted to standard as an American breed. It is the largest of them all, possibly the largest chicken in the world. It results from the crossing of eastern breeds, Black Javas, Dark Brahmas, Partridges, Cochins, with, possibly, the introduction of a bit of Mediterranean blood. This big chicken (the cock weighs thirteen and the hen ten pounds) has in its make-up got rid of much Oriental clumsiness, is quite a trim, active bird and a fair egg producer. It marks the direction which chicken breeding, still in its infancy, is taking.

QUESTIONS

1. What can you say of chickens as producers of wealth? How do they compare with wheat? With cattle?
2. What is the origin of the barn-yard chicken? Describe the wild ancestor of these chickens. Where is the jungle fowl to be found today?
3. Who first wrote of the chicken? How did it come to Europe.
4. How is it proved that chickens and jungle fowls are of the same tribe?
5. What is the "galliform order"? What are its peculiarities? What are some of the members of this order?
6. Most of the domestic animals were tamed in the long ago. What are those more recently tamed? What contributions did America make?
7. Have you ever dressed a chicken for dinner or watched the cook do so? What can you say of the crop? What purpose does it serve?
8. Describe the gizzard. What is it for? In what class of birds is it most highly developed? What other animals have gizzards? Give examples of the strength of the gizzard.
9. Flocks may be improved through selective breeding. Show what may be done in a flock of black chickens.
10. Give examples of varieties of chickens that have been developed by selective breeding. What are two great races of modern chickens?
11. Tell how the chickens of the Mediterranean, of the Far East, came into the western world. What are the good points of each?
12. Show how the chicken developed into the Mediterranean strain when it was brought west from India. Describe the Asiatic type and show how it is different.

13. In America the breeders tried to combine these good qualities by creating new varieties. What types of these chickens do you know? What are their characteristics? From what races do you suppose they came?

CHAPTER XI

THE BALD EAGLE

IT had been seventy-five years since a certain pioneer had come into this wooded glade and chosen it as the place at which to cut for himself a home from the wilderness.

A rather peculiar fact had to do with the selection of this home site. In a tall pine tree on the edge of the glade two great bald eagles had built a nest and were raising their young. Because of them it did not seem so lonesome.

The pioneer had died long ago. His children had matured, lived their lives, and gone the way of all men. Their children had come to manhood and had deserted the cabin in the woods and it had fallen into decay.

Yet in the old pine tree the eagle's nest still held its place and each year its owners, the same pair of these lords of the air that had nested there seventy-five years earlier, came back to raise a brood of their kind. Thus had they measured the span of their lives against that of man and shown themselves the master.

This bald eagle is none other than the bird whose imprint appears on the gold and silver coins of the United States, whose figure graces the top of the standards to which are attached the flags of the nation when its forces are on parade or going into battle. This bald eagle, in fact, is the American eagle, the bird which was long ago selected as the emblem of the United States.

The bald eagle is the king of the air. It is to members of the feathered world what the lion is to the furry fourfoots. It is the fighter that stands at the head of the list, the bird which, when it appears in Natureland, thrills all of its fellows round about as in olden days humble subjects were moved by the appearance of their king.

The eagle stands at the head of that order of birds which the scientists call the "raptores," or birds of prey. These birds of prey are a rather small group, compared with other orders, such as the "passeres," or perching birds, or the "pygopodes," or diving birds. There are but four bird families in the raptores order. Two of these work the day shift and two the night shift.

One of the day-shift families includes in reality no birds of prey, but creatures which feed upon the flesh not of animals that they have killed, but of those that have otherwise met their death. This is the family of vultures, that we call buzzards, which work as scavengers, and render yeoman service in sanitation, by devouring the bodies of dead animals wherever they are left exposed.

To the other family of the day shift belong those masters of the air, the eagles, and their smaller kindred, the hawks and the falcons. They are the tigers and wolves of the feathered world.

Though the bald eagle is called the king, he is no more without rivals than is the lion, the biggest of which beasts, in fact, are fifty pounds lighter than the biggest tigers. These latter used to win when the two were matched to fight in the arena at Rome. Even in the territory which is its special home, North America north of Mexico, the bald eagle has a rival, the golden eagle, whose size and power are not greatly different from its own. The bald eagle and the golden eagle, however, seem to have divided territory, and seldom come into conflict. The golden eagle is a creature of the mountains, taking its tribute among the wooded wastes, while the bald eagle haunts the lower levels and takes its prey along the rivers and the shore lines of oceans and lakes.

Scattered over all this territory where the bald eagle and the golden eagle hold sway, are the lesser barons, the hawks, spreading terror in their smaller way, but losing caste when one of the masters appears.

When America was young the bald eagle reigned as royally as any monarch the world has ever known. Through all of the territory which is now the United States, on through Canada, and blanketing Alaska to the Arctic, were to be found the mighty eagles, enthroned in their lofty nests in tall tree tops or on towering cliffs, each overlooking the realm it ruled.

Three feet to three and a half was this bird from the tip of its beak to the tip of its tail. Measuring from the tip of one wing to that of the other it was found to have a stretch of six feet, seven feet, possibly eight feet. Its weight ranged from six pounds to twelve pounds, which does not seem so great, and is, in fact, only about that of the larger barnyard chickens. It is, however, great among the birds in which the power of flight is highly developed, for it must be borne in mind that flying birds, like airplanes, are built on a plan combining strength and lightness.

This bald eagle is not bald at all from the standpoint of having no feathers on its head. It is called the bald eagle from the impression one gets of it at a distance. The head of the adult is pure white and this color extends well down on its neck so that its head looks bald. Its tail, also, is white, while the body is a dark, rich brown. Young eagles are brown all over, and do not get their white heads and tails until they are three or four years old. For a long time naturalists thought that these big, brown birds belonged to a different species from those with the white heads.

This huge bird that preys has through the generations itself been regarded as a prize for the huntsman. Whereever a great bald eagle has been found it is likely to have been killed if an opportunity offered. So it has come to pass that its ranks have been gradually thinned until today it is almost or never seen in settled regions where formerly it was abundant. Today one almost never sees a bald eagle along the rivers of the eastern states. Pennsylvania, for instance, which was once one of its favorite hunting grounds, now sees few of its kind. It has, in fact, been thinned out until it has almost ceased to exist in much of the United States.

In Alaska a splendid race of bald eagles, even bigger than those of the United States, is to be found. Even in that isolated territory, however, a price has been placed upon the bald eagle's head by the authorities, and, as a result, it has been slaughtered by the thousands. The great people which chose the bald eagle as their symbol is steadily exterminating it. Unless an understanding of what they are doing can be impressed upon our huntsmen, the time will soon come when the bald eagle will take its place beside the passenger pigeon as a splendid

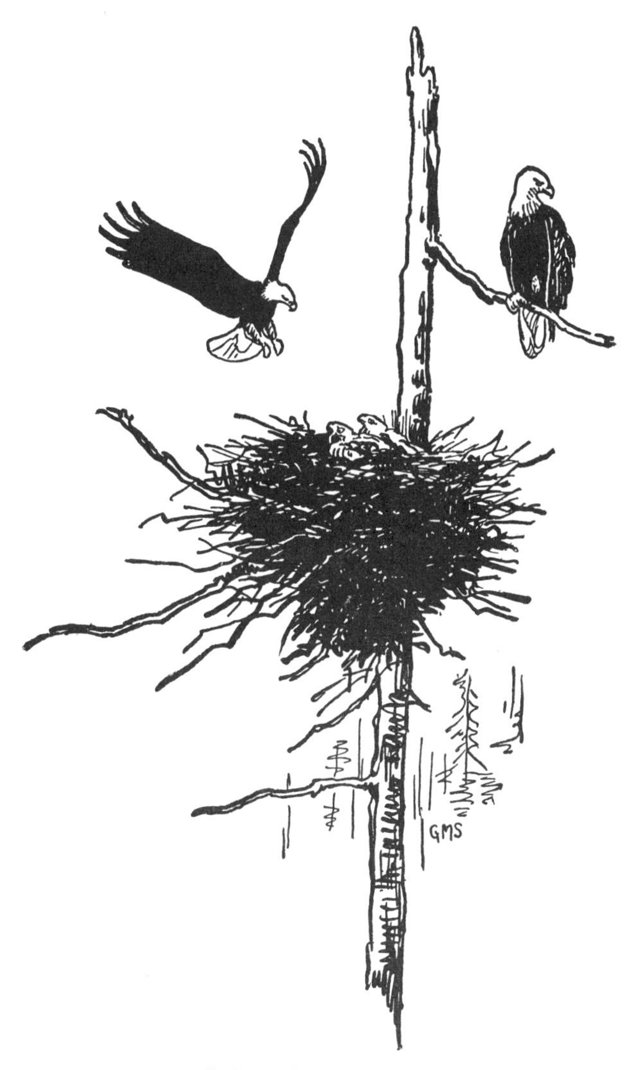

BALD EAGLE AND NEST

American species which once glorified its solitudes, but which no longer exists.

Imposing were those homes of the eagles in the shadow of which the early settlers built their cabins. Wise in their instinct the eagles always selected the largest trees of the forest in which to build their nests. Usually these towered scores of feet before their trunks put forth any limbs, and thus were difficult, if not impossible, for enemies to climb. So were the castles of the eagles protected as though by walls and moats and drawbridges. Often the eagle's nest was found in a towering pine that reached ninety feet before a limb was found. Then another thirty feet farther up the tree might branch out in such a way as to offer a crotch favorable to the eagle's plan of building.

Here he starts his nest. In this crotch he begins to assemble pieces of wood ranging from mere twigs to sticks three inches in diameter. The run of the timber for the eagle's nest is tree branches, as thick as the wrist of a man.

The nest that the eagle builds is five or six feet across. Before he is through he is likely to make it a substantial structure, four or five or six feet thick. The top of it is bowl-like in shape, and this he carpets with moss or grass or leaves, making a soft bed for the young ones that are to come. There in the top of the pine or other large tree this nest forms a massive blot that can be seen against the sky for miles around.

A single pair of birds may add to the old nest for several years, allowing the original foundation to stand. Finally, their instinct warns them that decay may be eating at it, so one season they set about tearing down the old house and building another one on its site, just as might a human family when the old homestead fell into decay. They wreck the old nest, throwing its rafters and beams overboard, and build another of new timber.

The mother eagle usually lays two, sometimes three, quite large eggs, perhaps three inches in length, bigger than those of the domestic goose or turkey.

It is during the nesting season, when the greedy, growing young ones are demanding food, that the sway of the bald eagle over its domain is most autocratic and severe. It is then that it must hunt with a ruthlessness that it need not show at other seasons.

This tall tree or rock ledge on which the eagle builds its nest is sure to be in the vicinity of a watercourse, or shore line, for the bald eagle depends very largely upon fish and water birds for its food. As is the way in the realm of nature, it has divided the field with the golden eagle, whose chief delight is grouse and rabbit, back in the woods and parks of the mountains.

It is early morning in eagle land and the family is to be fed. The mother eagle leaves the nest, wings her course to a wooded lake a dozen miles away and circles about its shore line on the chance that some fish has met misfortune in its depths and has been cast ashore by the waves, ready for her breakfast. Finding such a fish, she may bear it away to the nest to stay the hunger of her young ones until more daring hunting can furnish choicer food.

In the meantime the father eagle has sought a lookout post on a tall dead tree, from which may be seen a stretch of river winding for many miles. Here he sits and watches.

To this task he brings one of the best pairs of eyes in all

the world. There is no doubt that, just as the sense of smell among the insects is much more highly developed than anywhere else in nature, so is the sense of sight more highly developed among the birds. Men prize the sense of sight more highly than any other gift, but man's dependence upon it, his use of it through every moment of his active life, is by no means as great as that of the birds. The swallow pursuing an insect in its zigzag course through the air, the woodcock hurtling headlong among the branches, this eagle surveying his field for any chance victim, is even more dependent upon high quality vision than is man.

The eye of the bird when one looks at it gives the impression of great alertness, of clearness, of efficiency, and it has in truth all these qualities. It is bigger in proportion to the size of the bird's head than are the eyes of most of the other animals. It is a highly developed eye. It must, for instance, be able to make out objects at a great distance so that the bird may avoid enemies and catch prey. It must be no less efficient, however, in examining an object but an inch or two from it, for its food is close at hand when it must pass upon it. So it must be able to adjust itself to varying distances.

The eye of the bird has another advantage over that of the mammals in the fact that it has an additional lid. It has a curtain that it can pull over its window, a curtain that still lets in a good part of the light, just as do the white ones at our own windows. If the bird gets something in its eye it does not need to use the ordinary lid. It runs this film, through which it can still see dimly, over the eye to clean it. This film passes from the corner of the eye out and any experimenter can see it work by tickling a chicken's eye with the tip of a feather.

As the father eagle waits here at his lookout post there appears a speck way up the river course, a tiny moving thing in the distance which might be located by man's high-powered telescopes, but could never be seen by his unaided eye. Already, however, the eagle has made out the speck. Already he has noted the manner of its movement as it comes down the watercourse and formed his conclusions as to its nature, as to whether it is a duck, a goose, or a crane. He has made his decision as to whether he will try to overhaul it. Long before it has reached a point opposite his perch, if he has chosen to act, he has glided into the air and circled aloft.

It is a part of his strategy that this voyager, which now, even to the eye of man, has taken on the form of a mallard duck, should pass far beneath him. As it does so he starts in pursuit. This swift-winged duck, seeing the danger, plunges on at that terrific rate of speed of which, in fright, it is capable. This king of huntsmen must use his utmost energy, there overhead, to keep the pace set by the waterfowl.

But now he is close behind his prey and far above it. Then begins the final stroke of life or death. The eagle points his beak downward, and converts the height to which he has risen into an increased speed. On this down grade toward his prey he plunges headlong at a frightful speed. Now, possibly, he is coming twice as fast as a duck on the level can ever go. So in this final swoop of death does he have his victim at a disadvantage, and so, unless that victim is fortunate in dodging the approaching menace, its death is inevitable. The probabilities are that the talons of the eagle will sink home and that it will circle away to its nest with dark meat for breakfast.

BALD EAGLE CHASING A MALLARD DUCK

In the meantime the mother eagle has returned to her lake for fish. She has resorted, however, to a bold device of her kind for procuring choicer food than that which the waves throw up on the beach. She watches from aloft the work of that lesser bird of prey, the osprey, or fish hawk, from which she collects tribute. This osprey is circling over a quiet cove by the side of the lake, in which it knows, as well as any man fisherman accustomed to this body of water, just where fish are thickest. When the fish hawk sights its quarry in the placid waters of this cove, down it plunges, with a mighty splash, and seizes its prey.

With this choice fish in its talons the osprey takes wing, shakes the water from its feathers, and starts for its nest. The watchful mother eagle gives chase. She dives beneath the fish hawk, causing it to mount to ever greater heights. Carrying its burden, the fish hawk is at a great disadvantage, yet stubbornly it clings to it. Despite one charge after another from the great eagle it refuses to give up its breakfast.

Finally, out of patience, the eagle soars high above it, descends with a mighty swoop, talons ready for a stroke, beak open, screaming frightfully. In the face of this terrifying charge, the fish hawk loses courage, drops its burden, and speeds away. This is the opportunity of the eagle. She swoops at the falling fish, seizes it in the air long before it has reached the tree tops, and bears it away.

More thrilling even than this is the hunt in which both father and mother engage along toward evening. In this hunt they use teamwork. One of them swoops upon a duck on the water's surface, and that duck, appreciating its best chance for escape, dives beneath the surface. Quickly it comes up, realizing that its enemy needs some time to change its course and make a second swoop. In this attack, however, the second eagle strikes the moment the duck appears, and it must dive again. By the time it reappears, the first eagle has turned about and is ready to strike again. Thus by taking turns, one of the eagles is always ready to pounce the moment the head of the duck appears. Thus do they so exhaust it that soon it must remain on the surface long enough to recover its breath, and when it tries to do so it is seized by one of its enemies.

This manner of life of the great bald eagle very seldom in any way interferes with the activities of man. The bald eagle may occasionally bear away a new-born lamb or a small pig, but such occurrences are more accidental than otherwise. This great bird has even been accused of attacking small children and bearing them off, but it is doubtful if ever in the history of the world such a thing has actually taken place. It is known, for example, that few bald eagles are powerful enough to carry a weight greater than ten pounds. Only a very young baby is ever so light and these birds know their business too well to hunt for game which they could not carry. So it is certain that there is little reason why man should consider America's national bird his enemy and seek to wipe it out.

The campaign against the eagle in Alaska, experts agree, was not entirely justified by the facts. Salmon packers claimed that they consumed great quantities of salmon when the fish came into the rivers to spawn. Those who know the habits of both salmon and eagles say that most of the fish eaten by the birds are those that are already dead. After the salmon come in from the ocean and lay their eggs in the inland streams, Nature considers that they have run their course and they die. It is upon these dead salmon that the Alaskan eagle chiefly feeds during the salmon season. These facts alone never justified putting a price on the head of the bald eagle.

Man has always made a big mistake in the position he has assumed toward hawks and eagles. Because some few of them have preyed upon the chickens of his barnyard he has thought that all members of the family were his enemies. He has, therefore, declared war on the whole lot. Wherever he has had a chance, through the generations, he has shot hawks and killed them. He has passed the sentence of death without giving the accused a hearing, just as he has killed the bald eagles without knowing whether they were his friend or his foe.

As a matter of fact the majority of hawks are among man's best bird friends. Nearly all of them work all the time in his interest. There are about fifty kinds of hawks in the United States, and of the fifty only two or three are enemies of man and prey to any great extent on his poultry.

Take the most familiar types, for instance. There is the red-shouldered, medium-sized hawk, that is quite wide spread, which most people have heard utter its "kee-you, kee-you" scream, and which is commonly known as the chicken hawk. It is not a chicken hawk at all. It lives almost entirely on rats, mice, frogs, and the like.

Then there is the familiar red-tailed hawk, a little

larger than the red-shouldered, often encountered about the farm. Then comes the marsh hawk which flits low over the meadows looking for rodents and rarely comes near the barn-yard. There are the big, clumsy fellows, which seldom catch a chicken, but which do yeoman service in picking up rats, gophers, prairie dogs, rabbits. Finally there is that little fellow among the birds of prey, the sparrow hawk, which, the experts find, eats insect chiefly, mixed now and then with small mice and birds. Even these are on the black list of man and he kills them on sight.

All of them, however, work steadily in his behalf. Students of agriculture know that rodents are among the worst enemies of the farmer. Rats and mice destroy more grain in this world than does any other agency. The Department of Agriculture, in its bird court, has tried all of these hawks. It has examined the stomachs of great numbers of them to find out what they really eat. It has found that all the hawks but three eat rodents chiefly and chickens or wild birds only occasionally. These experts have estimated that one of these big hawks living about the fields of the farmer will, in the course of a year, kill rodents that would have destroyed a hundred dollars' worth of grain. That hawk has saved the farmer a hundred dollars. Even though in the course of the year it has carried away a half a dozen chickens, it is yet most beneficial to this farmer. It is in his interest to encourage and not to kill it.

The three villain hawks that deserve death are the cooper hawk, the goshawk, and the sharp-shinned hawk. These the farmer should shoot on sight. They are active, swift-flying hawks, that hang about the woodlands, hide themselves in the thick branches, dart forth with great speed, and seize the small chickens. Their habits are different from those of the hawks of the open. They are sly. They work under cover. They are not creatures of the open as are the other hawks.

Any farmer who stops to think of it knows these chicken-catching hawks from the rodent eaters. Once he stops to think he realizes that the hawks of the open are his friends. Nobody who has not thought out this situation, who does not know the real chicken hawks from the rodent catchers, has a right to point a gun at any hawk. Maybe he will kill the wrong bird, and if he does he will have done the farmer as much harm as though he had shot some of his stock. A hawk should never be shot except where it is known to be an enemy and not a friend. Shooting one's friends is bad practice.

QUESTIONS

1. The bald eagle is the bird that is taken to typify the United States. In what ways is it used as an emblem?
2. To what order of birds does the eagle belong? How many families of birds of prey are there? Name them. Describe the manner in which they seem to work in shifts.
3. Is the bald eagle king of the air? Compare it with the golden eagle. How do these two divide the hunting grounds? What was the situation as to the eagles before white men came to America?
4. What are the measurements of the bald eagle from tip to tail? Its wing spread? Its weight? Why is this bird called "bald eagle"? What is its color?
5. Are bald eagles growing scarce? How does this come about? What should be done about it? What is each person's duty in this connection? What mistake was made in Alaska?
6. What kind of tree does the eagle select for its nesting place? How big is the nest? Of what is it made? When is it rebuilt?

7. How many young eagles does a pair usually have? What are they fed? Show the part that each parent may take in feeding them.
8. Describe the father eagle's wait for a duck. The sharpness of his eyes. How do the eyes of birds compare with those of other animals? How does its extra eyelid work?
9. Tell of the pursuit and capture of the duck.
10. In the meantime the mother eagle has been hunting. Show how the osprey catches fish. How the eagles rob him. Describe the duck hunt in which the two parents join.
11. How does the bald eagle sometimes harm man? Does it carry off babies? Is the harm it does such as to warrant its being wiped out?
12. What is the great mistake man has made with respect to eagles and hawks? How many kinds of hawks are there in the United States? How many of them benefit man? How many injure him?
13. What is the principal food of the red-shouldered hawk? The red-tailed hawk? The marsh hawk? The sparrow hawk? Does their foraging help or hinder man? How much may a rodent-hunting hawk save a farmer in a year?
14. What are the injurious hawks? What is the rule that should apply to hawk shooting?

CHAPTER XII

THE HUMMING BIRD

THERE are humming birds so small that one of them, without its feathers, might easily be put into a thimble.

They are the smallest of all the birds of the world. Even the Lilliputians would have been disappointed had they been served one of them for dinner on Thanksgiving Day.

Yet no bird has a better development of breast meat in proportion to its size, for breast muscles are for the working of wings, and the humming bird has more vigorous wing movement than any of its larger fellows.

The humming bird, despite its small stature, is, in fact, one of the marvels of the bird world. It surpasses all the rest of them on many counts, and does many strange things which none of the rest of them can do.

It can stand still in the air, for instance, as can almost no other. It is by standing still in the air that it makes its living. A human being who could stand still in the air could make a living also by going into vaudeville. The humming bird, however, performs this feat as a part of the work of getting food just as man milks a cow or digs potatoes. It does the trick for its dinner.

The humming bird flies up in front of the flower on a trumpet vine, or a honeysuckle vine, or a jimpson weed out in the field—always selecting some flower with a very deep cup. It hovers immediately before it. It is hard to see just what it is doing, but here is what takes place.

Standing still before this flower the humming bird points its bill straight into it. Then it sounds the foreward gong and its engines drive it to the front just a little, perhaps an inch or an inch and a half. Its bill slowly finds its way down the channel that leads to the very bottom of the flower. There it encounters a very choice and sweet titbit—nectar. It laps it up, reverses its engines, and slowly backs up.

This backing up is in itself, from the bird standpoint, a very marvelous performance. The humming bird has all the rest of them beaten in this respect also. It is the only bird in the world that can fly backwards. It is otherwise remarkable in its performance on the wing. It darts with inconceivable swiftness. One instant it may be hovering by a flower here, and the next be forty feet away. Its quickness, its swiftness on the wing, is beyond measuring. Its control, its ability to dart sideways, to avoid hitting branches of trees is remarkable. Probably, on short dashes, it is the swiftest of living creatures. Certainly it is possessed of a rare grace and precision of movement.

There is a whole romance of the animal and vegetable world back of the visit of the humming bird to the trumpet flower. Certain of the trumpet vines profit very much by the visits of the humming birds; the very existence of their kind depends upon them. They offer rich bribes to the humming birds to come to see them. These bribes are in the form of that thing which humming birds like best—honey. This the flower has placed deep down in the bottom of its cup. Humming birds are fond of small insects also, and these may be found in the flowers, where they also, no doubt, are attracted by the honey.

HUMMING BIRD AT A TRUMPET VINE

Humming birds know that trumpet flowers afford their whole bill of fare, so they visit them eagerly. They stand still in the air, reach in with bills built for the very purpose, and help themselves. While they are doing so the trumpet flower sifts a fine flower dust called pollen onto the feathers of their heads. Putting this pollen on the heads of the humming birds is a matter of life and death to the trumpet vine.

The humming bird goes away with this pollen sticking to its head, and visits another flower for another bit of honey. Some of the pollen that has stuck to it in the first flower is shaken off in the second. Some of it reaches the very young seeds of the flower, and is necessary to their further growth and to the ripening of seeds. Without this transfer of pollen there would be no seeds to make other trumpet vines. It is because of it that the race of trumpet vines survives.

The way the pollen of most flowers is transferred from flower to flower is by being carried by bees and other insects. The bees do a greater part of the work of pollen carrying. There would be no apples, peaches, or cherries if the bees did not carry pollen. The humming bird does this sort of thing for the flowers of certain trumpet vines and other plants. They are plants with deep cups and store their honey in such a way that only the humming birds can get it.

Many insects, particularly the bees, vibrate their wings very rapidly. They can stand still in the air. The birds' scheme of flying is based on a different idea. They depend on soaring chiefly with but an occasional flap to keep them going. The humming bird alone buzzes its wings and all because of its habit of drinking from the flowers as do the insects.

This tiniest of birds can truthfully make another boast, a very large one in which it claims to surpass all others.

It can truthfully claim to be the most gorgeous, highly colored, gaily-trimmed, and beruffled, in fact the most beautiful, of all the members of the feathered kingdom. Since birds are, admittedly, the most beautiful of animals, the humming bird, it would seem, might quite properly lay claim to being the most beautiful creature in all the animal kingdom. It is often called the gem among birds, since it has borrowed the colors of all the precious stones. It is often likened to a fragment of the rainbow, so vivid and varied are those colors. Of all living things it is said to be the most elegant in form and brilliant in color. Naturalists who devote their lives to bird study agree that the humming birds far outstrip in beauty and variety their nearest competitors, the birds of paradise and sunbirds of the South Seas, or the peacocks of the barn-yard.

This tiniest and most gorgeous of birds belongs exclusively to the western hemisphere. Nowhere else is it to be found. The very center of its domain is the point where the Andes Mountains of South America cross the equator. It likes wooded mountain sides. They provide a luxuriant vegetation, an abundance of flowers that never fail, and are therefore to the liking of the humming birds. These birds are not abundant in the lowland forests of Brazil as people usually think. Peru and Colombia rank next to Equador in humming bird population. Then they extend up across the Isthmus of Panama, through Central America and Mexico.

Altogether in these tropical countries there are more than six hundred different kinds of humming birds. There are between fifteen and twenty species that come up the Pacific coast from Mexico in the summer time and make their nests in western United States. One or two even reach Alaska.

East of the Rockies the ruby-throat sweeps up with the spring in a gaudy host, spreads itself out over half a great nation, drives on even to Labrador, and thus gives nearly a hundred million people a chance to glimpse the miracle of it in their summer gardens. The ruby-throat is the greatest migratory humming bird, the traveler, the bold adventurer who voyages from the tropics well over the temperate New World.

The two elements in the outstanding beauty of the humming bird are its coloring and the fantastic arrangement of its plumage. Its dainty feathers are put on with an incomparable smoothness and over their surface runs a sheen, an iridescence, a changing of colors, of which an idea can be gathered from watching other birds, though none of them in this respect equals the humming bird. Its plumage is like the surface of highly polished metals, steel, gold, or copper. It borrows the coloring of precious stones, is as brilliantly green as the emerald, as radiantly red as the ruby, as richly yellow as the topaz. Its throat may be green, its crown black, its body scarlet, its tail running into yellows, and its darkly purple wings. All of these colors play hide and seek with each other, running the scale of the different shades as the light plays upon them. Each of the six hundred species has its own color scheme and designs. Aside from the few dull-colored species, each seemingly tries to outdo the others in the brilliant coloring of its dress.

No less fantastic are the patterns of the costumes that

each of these small birds wears. Take the racket-tailed humming bird, for instance. Its tail consists of ten feathers, two of which are twice as long as its body. They are merely quills, stripped of their barbs, until they get near the end, where they flare out into plumes. In addition, this wonderful bird wears about its legs great powder puffs of white down—as dainty a pair of boots as one could imagine.

The thorn-tailed humming bird has stiff quills for tail feathers and other quills shaped into a long, tufted peak on its head. The coquette has a brilliant ruff about her neck, two graceful feather horns, and a broad tail. Others have fan-like crests and fluted tufts of brilliant feathers used in fanciful designs for their adornment. In these humming birds Nature seems to have decided to forego all restraint and try her hand at creating the most radiant and fanciful creatures of which she is capable.

Another respect in which the humming bird surpasses all others is in the length of its bill in proportion to its size. This bill must reach to the bottom of deep flower cups and so is often very long, considering the small size of its owner. Most of these bills are straight, but the sickle-billed humming bird has one that curves in almost a half circle that it may get honey out of a flower with a corolla of that shape. The siphon-billed humming bird sometimes has a bill five inches long, longer than its own body.

Inside these bills are the strangest of bird tongues, which may be run out through their tips when needed. The long tongue is a tube, and the bird may, as the butterflies do with a similar tongue, suck honey up through it as one would drink lemonade through a straw.

Only a small proportion of the food of the humming bird, it should be remembered, is honey. By far the greater part of it is made up of tiny insects and spiders, which live on the inside of flowers. This long tongue has "glue" on it, as has that of the frog, and so, when it runs down for the honey, it likewise mops up many of these tiny insects and adds them to the humming bird's dinner.

Most of the humming birds are satisfied to spend their whole lives in these forests of the tropics where flowers always bloom and insects are abundant. The ruby-throated humming bird, however, which is well known throughout the eastern half of the United States, sees fit to go avoyaging. It spends its winters in Mexico and Central America and, when spring comes, moves en masse two thousand miles to the northward. There it gives an example to many nature lovers of just how all the members of its family to the south conduct themselves in the course of their family lives.

The humming bird, like as not, may select the vine over your veranda or the old apple tree in the garden, or the blackberry bush by the rail fence, as the place for building its nest. Then it will busy itself about the woodshed or the thickets, gathering cobwebs, for they are its principal building material. These cobwebs it binds very securely about a twig, and from them and such dainty materials as thistledown or dandelion down it builds up a cup-like nest, very firm and finished, and as soft inside as an eiderdown quilt. If it is on the apple tree branch it will cover the outside of it with lichens and make it look so like a knot that it would never be noticed. If it is in the blackberry bush among the snow-white blossoms, it will splotch it with white feathers that look for all the world like flowers. Altogether it is a wonderful structure that this tiny bird builds with no other tool than its long bill.

The inside of this nest is little bigger than a walnut shell, or at most, half of an egg shell. In it the mother bird places two small, white eggs, little larger than navy

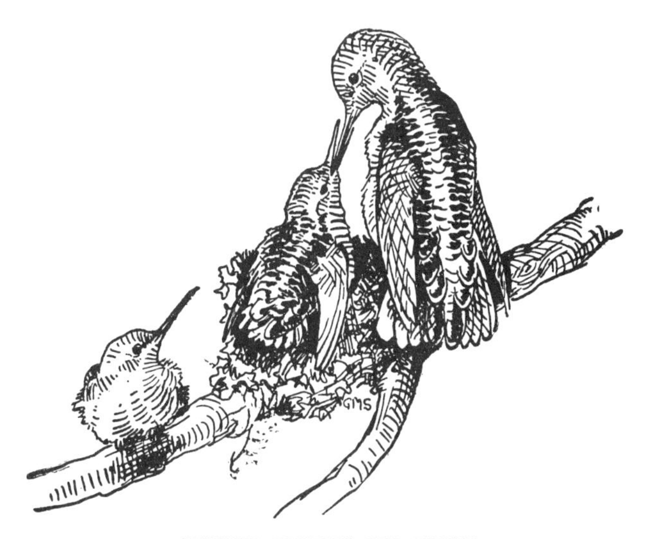

HUMMER FEEDING HER YOUNG

beans. From them in the course of time she hatches two little birds, so tiny and so without feathers that they remind one of nothing so much as honey bees which have been dipped into water and made to look very wet and forlorn.

For these ugly little ones the mother bird hunts busily,

and fills her tiny crop with honey and insects. When she comes to feed them she is likely to give any person who is spying upon her a great deal of a fright. One of these tiny, fragile, baby birds lifts its delicate head and the humming-bird mother adjusts her long bill to its open mouth. Then, horrible to relate, she drives this sword-like bill a full inch down the throat of the little one. It would seem impossible that she should not have driven it entirely through her baby. As a matter of fact, however, she has but put its tip down into the crop of the baby hummer and, having done so, she starts the pumps to working and soon the little one is full. Examine it after her visit and you can see its little crop distended. So thin and delicate is the skin of that crop that you can even see through it and find out what mother has brought for breakfast. Much of the food inside this crop is likely to be tiny spiders, some of them still alive and crawling.

The father humming bird has nothing at all to do with this family life of his mate. Long before the little birds have hatched he has left to his wife all the burden of taking care of the family; he serves only as a guard. For a few brief days at the time of the wooing he is very attentive to the mother bird and cuts fantastic capers for her entertainment. As she sits on a twig, for example, he flies into the air and swings back and forth in an arc about her, buzzing his wings at tremendous speed, and producing an odd burring noise as the wind whistles through his pinions. He pays constant court to his mate until the nest building begins. He seems to help for a while even in this work of nest making, but, gorgeous creature that he is, he grows tired of such commonplace drudgery, deserts the task, and flies away. The drab little mother finishes her nest alone, lays her eggs, sits on and hatches them without any attention from her mate. She must even leave her eggs and go hunting for her dinner.

Her mate does not share at all with her work of feeding the young ones, or teaching them to fly. During all this time he has been siting in solitary state in some tree near by, which he claims as his own. He seldom moves from his perch except when some other bird, be it robin, thrush, or oriole, trespasses upon his preserve. Then, tiny as he is, he darts out suddenly and gives battle to these birds many times his size. So swift is he and so sharp is his bill that he usually succeeds in driving them away.

Nearby this vigilant father there is likely to be some bush in bloom, such as the red-flowering currant, which he claims as his own, from which he gets his food, and which he defends as he does the tree on which he perches. If it should happen, during this nesting period, that his mate herself should approach his currant bush, he would fly at her most viciously and drive her away. Thus does it appear that the gorgeous humming bird is not a wonderful success as a husband.

One unusual thing about the members of the humming-bird family is the fact that they seem to have no fear of man. They will build their nests by the very side of his garden walk, or in the climbing rose that runs about his window. They will fly into his sun parlor and sip from the flowers that are blooming there. With a little coaxing the mother bird will light upon man's finger and feed her little ones as they rest in his palm.

Should the mother of these little ones meet misfortune, some human friend may take them into his house, feed them on honey, and thus bring them up. When they are grown birds, however, this honey will not be strong enough food to keep them for long, and in the end they will die unless they are given flowers from which to feed and in which they get their insect food.

The humming bird belongs to a family that is quite set off from most of the other members of the feathered kingdom, but is more nearly related to the chimney swift than to any other bird we know. As a matter of fact the humming bird is like the chimney swift in many respects. Its wings, for instance, are quite similar. The first and second joints of those wings are very short in proportion to the third joint, which is the important part of the humming bird's wing. The long pinions of this wing tip cross over its back as do those of the chimney swift.

The humming bird is like the chimney swift in one other respect—in the fact that its feet are so little used. Neither the chimney swift nor the humming bird alights on the ground and neither of them is capable of walking. The humming bird's feet are little more than tweezers, which fasten it to a branch on which it perches. If it wants to move ever so slightly, it lifts itself on its wings and sets itself down again.

QUESTIONS

1. Tell about a humming bird that you have seen. How small are they?
2. Describe their "standing still in the air." Why do they need to be able to do this? Why do they need to fly backward? Why are they swift on the wing?

3. How does the humming bird help the trumpet vine? How is it paid for the service? What animals carry most of the pollen for plants?
4. How does the humming bird compare with other birds in coloring? In beauty? It is said to be a gem among birds. What, in the animal kingdom, do you know that might be considered more beautiful?
5. Where are humming birds most abundant? Are they found in tropical Africa? How many kinds are there? How many in the United States? East of the Rockies?
6. Describe the sweep of the ruby-throat.
7. What can you say of the "sheen" of the humming bird? Of its coloring? Of the patterns of humming bird costumes?
8. Humming birds have very long bills. For what are they used? What is peculiar about their tongues? What do humming birds eat besides honey?
9. Where is the ruby-throat likely to nest when it comes north? What materials does it use in building? Describe the finished nest. How is it made hard to see? Give an idea of the size of this nest. Of the eggs. Of the young birds.
10. How does the mother hummer feed her young? Of what does this food consist?
11. What sort of person is the male hummer? Describe his wooing. His later desertion of his mate. Does he help feed the young?
12. Describe the life of the father bird in his solitary tree. His readiness to fight. His unkindness to his mate.
13. What is the humming bird's nearest relative? How is it like the chimney swift? In what respect are its feet peculiar?

Chapter XIII

THE OSTRICH

AS a humming bird would fit into a thimble, so might the carcass of an ostrich fit into an ash cart.

As the humming bird is the smallest of feathered creatures, so is the ostrich the largest. The biggest members of the ostrich tribe are about eight feet tall and weigh around three hundred pounds. They are the whales of the bird world.

The humming bird is one of the most beautiful of God's creatures, while the ostrich must be classed among the ugliest. The humming bird has its home among the jungles where vegetation is densest, while the ostrich seeks the solitudes of the treeless plains. The one flies with the swiftness of an arrow from the bow of an Indian, while the other cannot fly at all.

In all nature it would be hard to find two members of a single group more unlike, yet it must be remembered that hummers and ostriches both belong to the class Aves—both are birds. Far back in the past there may have been a bird-like creature that was ancestor to both of them.

If so, at some point their paths began to diverge, they began to live under different conditions, to be affected by different influences. The ancestors of the humming bird took to the forests, while those of the ostrich took to the open plains. The ancestors of the one took to sipping the sweets of flowers, while those of the other fell to cropping the grass of the prairies. The one, depending on limited quantities of choice food, grew always more refined, smaller, more delicate. The other, eating greater quantities of a coarser food, grew to a large size, developed a physique that fitted the conditions in which it lived.

The ostrich, living in the open country and cropping grass, found few occasions for flying, had little use for wings, but much use for legs. As the generations passed its legs grew stronger and its wings grew weaker. Finally it stopped flying altogether. Being unused, its wings grew weaker and weaker. The pinions on them lost their stiffness. The whole wing structure broke down. They became mere useless flappers, demonstrating the fact that unused members tend to disappear.

In the meantime the humming bird had found it necessary to hover in front of a flower, to stand still in the air, and the need resulted in its developing this knack which is possessed by no other bird.

It is the peculiar needs that have been thrust on different birds that have led to the great numbers of varieties of them. One took to the water and developed web feet to help it in swimming. So were the ducks evolved. Another went wading and its legs lengthened and it became a crane. Another needed a hook to help it in climbing and evolved the beak of the parrot.

The wing is not the only part of the ostrich which has demonstrated the principle of nature, the wasting away of unused parts. The ostrich originally had five toes like other birds. Today it has but two toes. It is the only bird with but two toes. Running about the plains it had no use for five toes, so three of them have disappeared. All the strength of them has tended to go into the third or middle toe. It has grown large and strong. The fourth toe has remained, but it is growing weaker all the time. In another hundred thousand years it also will probably have disappeared and the ostrich will be a one-toed bird.

It will not, however, be the first animal that did just this thing. The horse began as a dog-like animal with five toes. It, too, took to the plains where it used its feet for but one purpose, to carry it about. One toe developed until it became a hoof and all the others disappeared.

The ostrich is a dweller in the great open spaces of the world, where solitudes are most unbroken. It likes deserts and level plains, and shuns timbered areas and mountains. It has lived so long in such areas that its body and its manner of life have so adjusted themselves to the hard conditions there found that it has, despite them, been able to flourish and maintain its numbers.

Before man came to upset the balance, ostriches were very plentiful in the desert reaches of Arabia, Persia, and Palestine. Today, however, they are hardly to be found in all that area. They have long been abundant throughout much of Africa, however, ranging from the Sahara Desert on the north to the Cape of Good Hope on the south. It is in Africa that the ostrich has made its last stand, and though its numbers have been frightfully reduced in the last fifty years, many wild ostriches are still to be found there.

It is true, however, that the ostrich is a representative

of an early order of birds, just as the opossum is a representative of an earlier order of mammals. It is a representative of a group that has been tending to disappear from the world, just as the mastodon has disappeared. It is one of the most primitive of living birds.

As the opossum is more stupid than other races of mammals that have developed in more recent times, so is the ostrich more stupid, less highly developed than the other birds. Most of its relatives of earlier times have gone. Well preserved remains of birds twenty feet tall, and of their eggs fourteen inches long, have been found. The roc, mythical great bird of the east, is believed to have lived even in Marco Polo's time. This ostrich, in situations peculiarly favorable to it, has managed to survive.

On the plains of Africa, for instance, it has had certain advantages. It is a very tall animal and its head is set upon a flexible neck. There on the level, grassy plains it has been able to keep a sharp lookout for any possible enemies. Being a bird it has better eyesight than the mammals about it. Because of this it has been hard for the enemies of the ostrich to steal upon it. When an enemy did appear the ostrich would take to its sturdy legs and run away. Here again it has had a great advantage, for it can run faster than any other creature in the world. The horse is among nature's fleetest animals, and man goes forth to the chase mounted upon horses. No horse in the world, however, can hold the pace of a frightened ostrich when it starts careering across the plains. Lions, panthers, wolves, cannot run half so fast. What with sharpness of eyesight and fleetness of foot, this bird, though stupid, has been able to keep itself alive.

It is interesting to note the manner in which other animals that cannot see so well find a way to use the long neck and good eyes of the ostrich to their own advantage. Wherever a flock of ostriches is found in the African solitudes they are likely to be associated with such other plains animals as zebras and gnus. Whenever the os-

OSTRICH AND EGGS

triches grow uneasy and move away, these shorter animals that are not so well fitted to detect the presence of an enemy go along with them. Thus do the ostriches prove a great aid to other animals of the plains, serving as sentinels for them.

Another advantage that the ostrich has had in surviving in a modern world is that it starts a large family from one to three times a year. Thus it can lose the greater majority of its young ones and still maintain its numbers. The mother ostrich lays from ten to twenty eggs at nesting time. The male scoops out a great basin there on the plain and assumes the greater part of the family responsibility. The parents take turns at sitting and the care of the young. The male ostrich is a model husband and father.

This nesting time, however, is danger time, as the ostrich may not then so freely use its most effective legs to run away from its enemies as at other times. The hateful jackal is the worst enemy of the ostrich. It gives most of its time to these great birds at nesting time. It watches for a moment when the ostrich may leave its nest and then proceeds to secure for itself an abundant egg dinner. Observers say that the jackal will, with its nose, push an ostrich egg up on the rim of the nest and let it roll back into the nest and strike another egg. This will break one or both of the eggs and the jackal proceeds to its feast. When the young birds hatch, many of them fall victim to this marauder.

During the nesting season the ostrich defends its own. It is the possessor of a blind courage and some ability as a fighter. An angry ostrich was once known to meet death by charging the engine of a South African railway train. It battles by kicking. It kicks forward as does a man, but a blow of its great toe is not unlike the kick of a mule. It can send a hyena hurtling or knock a man down. When a man is attacked by an ostrich, however, there is safety in knowing one thing. The ostrich cannot kick an object that is close to the ground. By lying down one may escape the chief danger from an ostrich.

The past century has been full of tragedies in the animal world because during that century the man creature has completely occupied the world and has developed the high-powered rifle to wonderful efficiency. The ostrich has been greatly reduced in numbers and has faced extermination just as have the buffalo, the whale, the bear, the seal, and many other outstanding members of the animal kingdom. It probably would have been exterminated if it had not been domesticated.

During the past fifty years, however, the ostrich has been tamed. It has taken its place as one of that small group of sixty-odd animals which since time began have so yielded themselves to the influence of man as to be called domestic animals. The ostrich has written a new chapter in the history of man and his relation to other animals. It is the newest addition to that group that has become known as domestic animals. Today it is to be found on a multitude of farms throughout Africa and at occasional places elsewhere, as, for instance, in Arizona, California, and Florida, in the United States. There are probably a million ostriches in the world today that are being cared for by man, as are his cattle and his sheep.

The ostrich is thus made secure from extermination because of one thing—the beauty of the plumes which grow from its body and the use of those plumes for decorative purposes, particularly as trimming for hats. Even before the Christian era dawned the plumes of ostriches were used to beautify the trappings of kings. Rome in her glory knew the ostrich well, and there is a record of one of her wasteful emperors slaughtering six hundred of these stately birds that he might serve their brains for dinner. When the Crusades gave birth to knighthood and when it later came into flower, the ostrich plume became a symbol of rank. Throughout many centuries this bird of the desert furnished the mark that identified the plumed knight of the battle fields of the western world.

For many centuries it was the men of the race who created the market for ostrich plumes. Of late, however, the greater demand for them has been to trim the bonnets of women-kind. It was this demand that led the ostrich hunters during the last century to slaughter great numbers of wild birds, threatening their extinction, and during the following generation to turn to farming them and thus save them from being wiped out. It was this demand that led to such a trade in ostrich plumes that there is record of twenty tons of them, valued at $500,000, being sent in a single consignment in 1909 from Liverpool to New York on the steamship Mauretania.

On the wing of the ostrich there are some thirty-six plumes which are the choicest of the eighty or ninety that their bodies furnish. These plumes were undoubtedly once stiff quills used for flying. When the ostrich stopped flying, however, its quills tended to lose their stiffness. The barbs on the sides of those quills softened, and ceased to bind themselves together to resist the wind. They became fluffy and curly. They grew equally on each side of the quill and not unequally as on the pinions of flying birds. They became a mere fluffy protection from the sun for the body of the ostrich. Nature, as is her way, sought, however, to give beauty to these plumes. Particularly handsome are those of the male birds in their striking black and white. They have been styled "the most perfect decorative item in Nature's storehouse." Few products of Nature are handsomer and few better fitted to be preserved for long periods of time and conveniently used by man.

It was an odd thing that, through the centuries during which plumes were in demand, when ostriches were hunted for their feathers, nobody seemed to think of domesticating the bird and raising plumes as a business. The discovery of the possibilities that lay in this idea came about by accident. Back in the eighteen sixties a Kafir chief presented six pair of ostriches to Sir William Curry who was commandant of mounted police in Cape Colony, South Africa. Out of courtesy the commandant had to keep the birds and take care of them. They became gentle, were easy to handle, and multiplied rapidly. The idea of plucking their feathers and selling them and yet keeping the birds to grow other feathers naturally presented itself. Methods of harvesting the feather crop were developed. It was found that a bird with "ripe" feathers could be driven into a chute, that a bag could be put over its head, and that the plumes could be clipped off close to its skin. There was no pain to the bird in this and not a drop of blood shed. The plucking of ostriches everywhere is based on some variation of this simple method. Soon South Africa had many farmers embarked upon a new industry.

An ostrich, it was found, ate about as much as did a sheep and much the same food. A clover field was its delight, but chopped hay, bran, or grain pleased it when green food was not to be had. It would, however, eat

AN OSTRICH SWALLOWING ORANGES

almost anything that presented itself. It has a huge gizzard and needs big stones to help in the grinding. It is quite fond of broken rock such as is used in surfacing roads. Tame ostriches will eat pennies as fast as offered to them, and pick at buttons on the coats of visitors. Instinct tells them that these are of material that might help in grinding. Such ostriches have been known to pick watches out of pockets and swallow them. Sometimes they swallow objects that do them harm. One is known to have died from eating barbed wire. In California they are fed unmarketable oranges which they swallow whole and the course of which can be plainly followed as they travel down their long necks.

Unfortunately the frames of ostriches are bony and offer little meat and so the ostrich farmer may not sell ostrich steaks as a by-product to feather raising. There are possibilities in ostrich eggs, however, which are as good to eat as are those of any barn-yard fowl. The trouble is that they come in packages that are not convenient for using. One ostrich egg weighs three pounds and contains as much food as two dozen or more hens' eggs. There are few families large enough to serve even one of these eggs as an omelet. When so served, however, the quality of the omelet is excellent. If boiled they must be kept cooking forty minutes to be done at the center.

The ostrich has several relatives in different parts of the world, large, flightless birds of the open that have come down from the animal life of an earlier world. There is the emu, for instance, national bird of Australia, five to seven feet tall, with wings so small that one must search among its feathers to find them. It has hair-like plumage of no particular beauty and three toes. While the ostrich kicks forward, the emu operates differently and strikes out to the rear.

The cassowary is a six-foot bird of New Guinea and the Malay Archipeligo whose skin is there used for door mats. This bird has a helmet and the skin of its head is highly colored like that of a turkey except that it runs to brilliant blues, greens, and yellows.

On the pampas of South America dwell the rheas, western hemisphere representatives of these huge, flightless birds. They are about five feet tall, have gray feathers tipped with white, much used for making feather dusters, and their skins have been in demand. This fact came near leading to their extermination at one time. As many as four hundred thousand of them were killed in a single year. They are easy to tame and like living in pastures with cattle. On many of the ranches of the Argentine today one encounters flocks of these impressive rheas stalking about in solemn dignity. Thus will they probably be preserved.

QUESTIONS

1. Contrast the ostrich and the humming bird.
2. The ancestors of the ostrich were, at one time, probably not unlike other birds. Show how living in the desert caused them to change.
3. Show how the wings changed. What other bird have we studied which, leading an entirely different life, lost the power of flight?
4. The ancestors of the horse had five toes all but one of which were lost after it became a plains animal. Show what has happened to the foot of the ostrich through leading a similar life.
5. What continent is the chief home of the ostrich? It is of a very old order of animals. Compare it with the opossum. How does it rank in intelligence? Is its race increasing?
6. What advantages over other animals does the ostrich have on the plains of Africa? What other animals use ostriches in avoiding danger, and how?

7. What can you say about the size of ostrich families? What does this show as to the dangers which they face? How does the jackal plague the big bird? How does the ostrich fight?
8. The development of the repeating rifle has doomed many breeds of animals. How was the ostrich saved? Where are ostriches kept in pasture today? How many?
9. Tell the story of the ostrich plume. What created the modern demand for ostrich plumes? How does the ostrich plume rank in beauty?
10. Explain how the domestication of the ostrich came about by accident. Tell how ostrich farming grew to be a business.
11. In California people amuse themselves by feeding whole oranges to ostriches and watching them move slowly down their long necks. What other strange objects have they been known to eat? What is their chief food?
12. Is ostrich flesh good to eat? How big an omelet would one ostrich egg make?
13. What relative has the ostrich in Australia? Describe it. What relative has it in America? Describe it.

CHAPTER XIV

THE HOUSE WREN

THE riot call of the birds had been sounded in the back yard and all the tree inhabitants of the neighborhood had gathered. Why should they not respond, for was there not going on that most interesting of events, a fight between two of their kind? And had they not often observed that most superior man creature, and did not he always rush to the point where a fight was in progress?

As it is often the smallest of a group of boys that is set fighting for the amusement of the others, so, on this occasion, one of the tiniest of birds was at it. The house wren and the bluebird were having a tussle.

There was a feud of long standing between these two. The owner of this garden liked birds and understood them. He especially liked wrens and bluebirds. He had found a way to favor the wrens against the other birds, such as those noisy and greedy fellows, the English sparrows. If he made the doors of his bird houses no bigger than a quarter of a dollar, this man found, the wrens could get into them, but the English sparrows and the other bigger birds could not. So he made these doors by boring a hole with an inch auger and the tiny birds could get inside and defy the bigger ones.

But the wrens were not satisfied with this. They also wanted to fill up the bluebirds' houses with the big sticks they are so fond of using in nest building. The bluebirds had not all gone South for the winter, but some of them had hidden in the cedar trees all winter, finding there shelter

BLUEBIRD AND HOUSE WREN

from the cold, and berries for food. So they were on the ground early to occupy their old homes. They were the first birds to do nest building in the spring, and having occupied them, were not afraid to fight in their defense, despite the gentle beauty of their song and their reputation for happiness.

But the tiny wrens, arriving from the South a little later, were active and restless. They, too, liked bird houses. They never paused for a moment in their examination of all the possible homes in the neighborhood, and insisted on bringing sticks into many of them. They kept the bluebirds on pins and needles.

The bluebird that had gotten into this fight had done nothing more than defend his own threshold when down upon him had come this wasp-like wren, picking at his eyes, scolding in a most ungentlemanly way. The bluebird had resented the attack, had turned upon his attacker, and, being bigger, had driven him away. The wren had flown in among the thick rosebushes. This was his place of safety, for he was smaller and quicker among the briars than most birds. Safe among the branches he scolded the bluebird soundly, bursting occasionally into a mocking song.

The bluebird went about his business, but when he once more approached the bird house, that pest of a wren was again upon him. There was active darting and chasing of lightning quickness about the yard, an angry clatter of bird voices, ending again in the rosebush. So the fight went on, round by round, until the parties to it were tired out, were forced to give it up, without a decision as to which was winner, and the onlookers went away.

It is not unlikely that the wren's next combat will be with that brunette lady called Phoebe, because of the note she keeps singing, in the meantime marking time with odd beats of her tail. The Phoebe detests this ever-flitting wren and chases him under the house at every opportunity and then sits on the sill, snapping her bill threateningly. Even the song sparrow takes a fling at the wren, drives him into a bush, and sits on a tip of it and holds him prisoner.

A man observer of the contest between the wren and the bluebird would have thought that surely the former had been fighting for the very life of his young ones. Investigating, he would have found that the wren was not trying to establish a home at all, but merely a mock nest, a sort of playhouse that he would fix up just to amuse himself. He had been fighting, as a matter of fact, for the mere fun of it, and not that the nest was of any value to him.

These house wrens do this sort of thing all the time. It would seem that they are among the most active of all the creatures under the sun and do many things that appear to have no purpose other than to use up their energy and fill in the time.

Two male house wrens, for instance, carry on mock battles with each other that last for hours, simply that they may have something to do. They start by staging a singing contest. They sit on posts not ten feet apart and carol for an hour at the very top of their voices. Then one darts at the other and they circle and twist in and out among all the obstacles in the garden, all the time going at a frightful rate of speed and screaming noisily. One might think that this was a most desperate struggle, but if he watches carefully he will see that neither bird ever overtakes the other, that not a feather of the coat of either is ever ruffled. Sometimes, however, these combats are more serious. One Johnny Wren has been known to actually kill another. Sometimes, also, they break the eggs or kill the young of other birds. There are those, in fact, who feel very strongly against the house wrens because they annoy other attractive birds until they may be driven away.

It is all a part of the intense life that the wren lives, probably the most intense of all the birds. Birds live more intensely, it should be remembered, than do the mammals or any other animals. Their hearts, for instance, beat very rapidly, and their blood keeps up to a temperature of one hundred six to one hundred ten degrees, while man's is content to go along at about ninety-eight, and the reptiles, the cold-blooded animals, are comfortable with their blood ranging away down into the sixties.

The first one hears of the house wren is likely to be at daybreak of a spring morning late in April, when one awakes to a consciousness of a new note of joy in the air. The father wren has arrived ahead of his mate and is at his song. Ten times a minute he repeats it and his working day is no less than ten hours. The wren song is likely to be sung 6,000 times a day by a single bird.

It is a good song, for this bird is a member of the family of prize singers, a cousin of the mocking bird and the brown thrasher. Its song is not quite up to theirs, for it is not so big as they, but it is a good song at that, a rollicking song, very full of joy and happiness, energy and frolic, sparkle and cheerfulness. The cup of the wren overflows. It bubbles with enthusiasm. It keeps exploding in song. It is the bird busybody. It is about the liveliest creature in all the world.

It has a surprisingly big head for a bird of its size, a head with a sharp, business-like bill, and a pert, wide-awake look. Its body tapers rapidly toward the rear. It is gray underneath and reddish brown on top with bars across that become more pronounced the farther back they go.

The male bird takes a lot of trouble about selecting a nest before his mate arrives. He wants to have things comfortable for her when she comes. He may decide, for example, that the broken teapot which the mistress has put under the eaves of the front balcony is just the place to raise a brood. What a fuss he does make over his mate when she comes! He fairly quivers with joy at her arrival. Then he takes her to inspect the new home. And she, priggish person that she is, turns up her nose at his choice of a cottage and selects one of her own in the bird house at the end of the garden.

These nests are built in many odd places. The lady of the house may turn the sprinkling pot upside down to drain in the crotch of a tree. When she comes back for it she may find it very full of sticks and may hear the wrens quarreling noisily at her approach. Or the gardener may hang his coat on a fence paling and return to find a nest in the pocket. Wrens often insist, foolishly, on making their nests in the pump, while a cracked water jar in the tool shed is much to their liking. That space between the window blind and the wall is excellent, and a new roof put on over an old one is likely to furnish just the crevice for them.

In making these nests the wrens resort always to a certain bit of strategy which has back of it the fact that they are so small. They narrow the entrance of their nests until almost no other bird can get into them. The cactus wren of the Southwest, for instance, builds in the brush a nest that is a hollow globe, presenting thorns on all sides, with but a tiny tunnel leading to its feather-lined interior.

HOUSE WREN AND TWIG

The house wrens are among that small group of birds, including the English sparrow and the chimney swift, that seem to profit from association with man. The winter wren, a sweeter singer, may nest in the solitudes of the north woods and consider a journey to the United States as a migration South, may like the wild, waste spaces, but the house wren sticks close to man.

It finds that it is much benefited by doing so. There are, about the buildings of man, many choice nesting places, better than the mere knot holes and hollow limbs it used before he came. Most of the bird enemies, it finds, are afraid of man, and will not pursue it into his very presence. He is a protection. The wrens have lived so long with man that they have come to understand him quite well. There are those who believe that wrens actually come to human beings and make a great fuss to attract their attention when a young bird has tumbled out of the nest, or some other situation that they have not been able to meet, has arisen.

Certain it is that man and the wrens are on a basis of friendship that is quite satisfactory to both, though this fact in no way prevents Jenny Wren from scolding most severely if a human being gets too inquisitive about her premises.

Apple blossom time is love time for wrens. It is then that they are busy with nest making and pour out their song all day long. Jenny Wren, coming of such an industrious family, could not be satisfied with a family with a mere two or three children. She lays six, sometimes as many as nine, eggs.

Now she is busy with the sitting, but Johnny Wren must find some way to work off his extra energy. To this end he is likely to play busily at the game of nest making. He will choose some nesting place, maybe a bird house, and will fill it very full of sticks. In the midst of this he will fashion a very choice nest. He has been seen actually to bring many insects to these nests with which to feed the children of his fancy. He may build more than one of these nests before Jenny hatches the brood and gives him a chance to be of real use in this business of feeding the family, a chance of which he makes little.

And what a busy time that is for Jenny, six or eight growing children to feed and such appetites as growing children do have! The rate at which these young wrens develop is most surprising. It is only two weeks after they come out of the egg before they take to the wing. The wren parents working in the garden to feed their young are as good as a poison spray in ridding it of the insects that do it so much harm.

When this first brood is getting nearly grown, Johnny Wren, if the two stick together, induces Jenny to go over to the first nest site, the teapot, which he selected when he first arrived from the South. A second brood is to be reared and he argues that this is an ideal place for it. Jenny listens to his pleas, appears to be on the verge of yielding. Then she finds a knot hole through which she can get into the hollow of the wall and there she starts the second nest.

But here is another thing that not infrequently happens, Johnny, despite the fact that Jenny paid so little attention to his choice of a home, worked busily and sang happily through apple blossom time. Then, when sitting time came, he disappeared. The young ones hatched out were to be fed, and yet the father was absent. One might have imagined that some accident had befallen, might naturally think of the treacherous house cat. The facts were, however, that Johnny was bored with the prospect of too much family life, that, despite his seeming love of work, he was often a ne'er-do-well, an idler. He had, as a matter of fact, merely run away to the nearest thicket, and there, as early as four o'clock in the morning, could be heard putting on his concert to the dawn. When Jenny was ready for a start at the second brood he might be back, however, arguing with her over the nest site.

Ornithologists wondered for generations as to whether or not it was the same pair of wrens that came back to the same spot, year after year, to raise their broods. They wondered if the same Jenny held to the same Johnny as mate in rearing all the broods she mothered. Wrens looked so much alike and changed their coats between visits, so it was hard to say.

During the last few years a scheme has been worked out that, as time goes by, will solve many of the riddles of bird life, of their coming and going, of their routes of travel, of the length of their lives, of their mating. The scheme is based on banding the birds—placing little metal strips with numbers on them around their legs. A card with the record of the history of the bird bearing the given number is then filed with the Biological Survey, in Washington. Naturally, to keep the records from getting all mixed up, one office must keep them and reports must be made to that office. Whoever kills or captures a banded bird should write to "Biological Survey, Washington, D. C." and give information.

Much work has been done with ducks and other game birds. They are trapped at their breeding grounds in the north, and hunters, killing them here and there over the far-flung South, report the time and place. Thus can the extent of their travels and their migration routes be shown.

With birds like the house wren the method is quite different. The bander traps them in the bird houses that he has arranged about his place as homes for them. He finds one pair in the spring, for instance, living in a certain bird house and raising a brood. He wants to know if the same birds come back to the same nest for the summer brood, and if they come back the next year, after making the migration to the South. He puts a band with a number on it on the leg of each of these wrens. He may do this same thing to twenty pairs of wrens nesting in as many bird houses near his home.

When these wrens are raising the second brood he traps them all over again and makes a record of the birds that are mated and the bird houses in which they are nesting. The very first season of wren banding established one fact that had not been known before. Wrens do not stick to the same mates throughout their lives. They do not even stick to the same mates for one season. Wren society is full of divorce and remarriage. The pairs that are mated for the early brood are very likely to be broken up and the individual birds mated with others in the group for the second brood.

It has long been supposed that the same birds returned to the same spots in the North after their winter migrations. Banding these wrens proved this supposition to be a fact. The same wrens are trapped in nests in the same yard, mated with different members of the same group, year after year. The young birds raised by this group, it was found, are likely to return to the same general neighborhood, but not to the very spot of their birth.

These records of banded birds have not yet been kept for a sufficiently long time to learn many of the bird secrets. The fact of how long wild birds live, for instance, has not been shown. Many birds are being banded before they leave the nests in which they were born, however, and are being afterwards recaptured, and it is thus being shown that they are four, six, or eight years old, as the case may be. Soon it will come to pass that birds of the various breeds will be competing with each other to see whose life is longest. In this banding a method of finding the fact is being laid down such as never existed before. The opportunity for it is greater in the United States than anywhere else in the world because that country spans a continent in the belt of very busy bird travel.

The house wren is a summer resident in the greater part of the United States. There is, however, another bird much like it, called the winter wren, which chiefly breeds far up in Canada and to which the United States is a winter home. Again, there is the Carolina wren, a wondrously sweet singer, which is a more southernly bird than the house wren, rarely going farther north than Philadelphia to nest. Then again, there are such specialized birds as the shy marsh wren, with its long bill, whose nest is hidden among the cattails, and which never appears elsewhere. All of these birds are cheerful, busy, irrepressible little bodies, that help to make the world a very pleasant place.

QUESTIONS

1. Describe the feud between the wren and the bluebird. What was the cause of it all? What conclusion do you reach as to the nature of the house wren?
2. Dr. Robert Ridgway calls the wren a "feathered dynamo." Why? What facts prove that birds live more intense lives than other animals?
3. When, in the spring, does the wren arrive? What is its song like? How is the nesting place selected? What is a likely place for a wren nest?
4. How do wrens keep other birds from getting into their nests? Why do they want to be near houses?
5. How many eggs are laid in a wren's nest? While Jenny Wren is sitting, how does Johnny amuse himself? Describe his mock nests.
6. How do wrens help man to keep his garden in good shape? How does Johnny try a second time to select the nesting place?
7. Johnny may refuse to help support his family. What becomes of many father wrens after the sitting begins? Describe their life in the thicket.
8. How is bird banding done? What bureau of the government keeps the records? What is shown, for instance, about the travels of ducks?
9. How were the studies of wrens made? How was it shown that Jenny and Johnny believe in divorce?
10. How was it shown that birds come back to the same place year after year? Where are young birds, born in a certain back yard, likely to be found the next year?
11. How does bird banding make it possible to tell how long birds live? How does the United States offer the best chance in the world for this sort of study?
12. How does the life of the winter wren differ from that of the house wren? The Carolina wren? The marsh wren?

Chapter XV

THE HERON AND THE STORK

THESE cousins, the great blue heron of America and the white stork of the Old World, are quite distinguished birds, each occupying an outstanding, though quite different, place in the hearts of the people who know them.

The heron, America's master wader, is the melancholy sentinel in waste places, a creature of the solitudes, while the stork, fairy-tale bird of the children of Europe, said to bring the babies, second in their hearts only to Santa Claus, nests on the very house tops and hovers always near man as a good spirit watching over him.

It is not a big group which the great blue heron heads, but an interesting one with the night herons, or "squawks," and the bitterns adding to the numbers. The egret, unfortunate bird that has been done near to death that women might have its plumes for their hats; the lesser cranes of the swamps; the comic adjutant stork of Africa, five feet tall; and the jabiru of South America, fully as large, with a great bill a foot long, are striking members of the heron tribe, the stork-like birds.

One finds the heron where it is wet, where the turtle splashes, where snakes glide noiselessly through the slime, where the mellow voices of bullfrogs are heard in the twilight, where insects fill the air, where the foot of man comes not often nor easily. There he stands in a foot of water, alone in the world of waste, looking wondrously solemn, the very picture of gloom, yet, withal, assuming an air of dignity, almost majesty—a grand and stately figure in colors borrowed from the sky above and the water below—an art creation worthy of a great master.

One might think this great bird, four feet tall, a statue, so motionless is he. Then suddenly he springs into life, into action. He strikes. And in so doing he deals out certain death. That six-inch, dagger-like bill has run quite through some fish that had been idling in the shallows. It was for this the heron has been waiting.

This shy creature of the lonesome spaces is no mean huntsman. There are many morsels here in the water that are exactly to his liking. There is the water snake, that is to him what spaghetti is to the Italian. Any small member of the finny tribe, swimming within a foot of the surface, gives him his fish course. Any sort of a frog makes a plump sinker as a substantial part of his meal. A water rat lends variety. Crawfishes add a bit of roughness. Chance grasshoppers are like nuts after dinner. He fares sumptuously in a region that is difficult hunting for most animals.

It is for these items of food that he is here. It is for them that, through the centuries, he has been developing long legs for wading, and a long neck to give play to his javelin-like bill.

And here is an odd thing. When the heron stands there, so stately and picturesque, head up, seemingly unconcerned, his eyes are all the time looking into the water at his feet. They are so placed that they can do this. His head is broad at the top and these eyes are set on the sides of it, somewhat as though they were on the

GREAT BLUE HERON DEFENDING ITSELF AGAINST A HAWK

underside of an overhanging shelf. Thus placed they can keep watch in the water all the time.

It is odd how the manner in which the eyes are placed in the head gives an idea of the chief concern of their owners. In the beasts of prey, for instance, the eyes are in the front of the head, for they are hunters, their chief interest is catching some other creature that they may eat it. In the shy beasts that are preyed upon, as, for instance, the deer, the eyes are set at the sides of the head, are stuck out on pegs that they may see all about them. Their chief worry is a fear of being caught by these beasts of prey. Their eyes are lookout stations. Now here is a huntsman which is interested in game down at its feet and its eyes are fitted for looking downward.

It is surprising to find that the great blue heron, frail, slender, beautiful creature that it is, can successfully defend itself against the attacks of such birds of prey as the hawks and eagles.

While it is standing guard in the swamp, that competent huntsman, the red-shouldered hawk, may appear on the scene. It may see this awkward wader at its post. It knows that the heron's spare body affords little meat, but it is hungry and a bone is better than no food. It circles and swoops at its prospective prey.

The timid heron is much frightened. It knows, however, that it has no chance of escape in its slow and labored flight. Instead of attempting this it cowers on its haunches. It huddles down, all its plumes drawn close.

The hawk comes hurtling on. As it nears the heron it

finds that the wader is sitting tight with that long, needle-like bill pointed straight at its onrushing enemy. It is as if a man held a sword so that his foe might rush upon it.

The wary hawk appreciates this danger. He attempts to get past the protruding lance and give his talons a chance to drive home. Yet he finds himself quite unable to do this. He is forced to turn aside and forego striking.

These game birds are persistent, however, and the hawk turns, circles about, strikes again. At each approach, however, he finds that the heron is watchful, that its spear is so shifted that the attacker must turn aside or meet it head on.

The hawk, being bold, is likely to try a stroke, despite the danger. It is quite sure to receive a wound, though this may be but in the flesh. Feathers may fly. Blood may drip. The hovering heron is likely, however, to go unscathed. Its "present bayonets" has been effective.

If one intrudes upon the solitude of the heron it still does not forget the rôle of dignity and solemnity it is playing. Its rights as presiding genius of the lonesome marshes have been disturbed. Very well, it will go away from this place which is becoming crowded. It unfurls its great cinnamon-brown wings, six feet across, strokes the air, slowly, gracefully, noiselessly, and floats off in lofty and leisurely flight to some point where it will not be disturbed, its long legs dangling oddly as it goes.

Yet this is but one phase of the heron's life. He is a solitary hermit out in the marshes, playing a lone hand, departing whenever he is disturbed. This is because of a business principle held to by the heron tribe. Many birds hunt in flocks, but the principle of the herons is that they should parcel out the marshes and that each should have a roomy hunting ground all to himself.

When the day's hunting is over, however, this solitary bird wings its way to the heron village, maybe twenty miles away, where all is bustle, clamor, excitement. The herons in their family life are very gregarious birds, like the company of their fellows, and are given to chatter and noise. They live in colonies, hundreds of homes packed in a few score of trees, in an area of three or four acres.

They choose these homes with much wisdom. They seem able to pick the spot in a swamp that is surrounded by most underbrush and briars, and that is therefore hardest for animals that walk on the ground to reach. Or, perhaps, they may settle upon a clump of great fir trees in some such lonesome forest as that along the Columbia River of Oregon. In either case they add to the difficulties of getting at them by placing their nests in the very top of the tallest of the trees at hand. In the swamps the trees selected are likely to be great sycamores, and the nests, each as big as a washtub, will be placed at the tips of the branches that reach highest, maybe a dozen of them in a single tree. In the fir trees it is likely to be a good one hundred and thirty feet to the heron's nest and there it is comparatively safe from being disturbed. These long-legged waders appear oddly out of place perched in the tops of the great trees.

The nests are brushy structures, little more than flat platforms on which the eggs are laid. They grow bigger year by year and, in the sycamores in winter, make great dark spots against the sky. So large are some of them that, should they tumble to the ground, a man would have difficulty in lifting them.

The black-crowned night herons, lesser birds with much shorter legs and necks, are likely to occupy the smaller trees, the willows and the maples, in the heron colony. They are the night shift. To see them at the moment of their greatest activity one must visit the colony at dawn when the fishermen are returning from their work. The young herons in the nests keep up a clamor that suggests a swamp full of frogs. The parent birds squawk noisily. For the night herons this meal, served at dawn, probably should be styled supper, as it comes at the end of the working period and at the beginning of slumber time.

The night herons and the great blue herons, working different shifts, provide much the same food for their young—frogs, snakes, and fishes, and disgorge it in the same way down hungry throats. Both are a part of the same sort of noisy, harsh home life high up in the tree tops.

The night herons are still plentiful throughout America. Colonies of them may be found on Long Island, not far from New York City. The ranks of the great blue heron, however, are being steadily thinned. The time seems approaching when there will be no more of this bird, the prince of its tribe of long-legged fellows in feathers.

If the great blue heron is passing, if it should cease to exist in the world, there would be no upset in the sterner affairs of life, and few people would notice its passing. To be sure it serves a bit of useful purpose by eating snakes and field mice, and it is open to a bit of criticism because of the fish which it devours and which might be useful to man. The good it does and the harm it does,

WHITE STORK AND NESTS IN CHIMNEY TOPS

however, are of little consequence. Thin and bony creature that it is, it has no food value.

There is but one point of importance in thinking of the great blue heron and that is the peculiar element of beauty which it contributes to the world. The imagination of the artist is unable to conceive a bit of life which fits so ideally into a marshy landscape as the stately heron, standing majestically on one reed-like leg. A solitary sky view is never better broken by a touch of life than when the great blue heron flaps majestically across it.

The heron should be preserved because it is a thing of beauty, because it is harmless, because it serves no purpose when slaughtered but that of satisfying some hunter's desire to kill. It would be difficult to conceive of any set of circumstances that would justify any hunter in shooting one of these melancholy sentinels of the solitudes. Doing so is an act of folly which should not be tolerated by an intelligent people.

Then there is the white stork, prince of waders of the Old World, one of the birds which, through the centuries, has had a rarely equaled place in the affections of the people, and which, it may be, is the best loved bird among the peoples of Western Europe, Asia Minor, and the Mediterranean area. The stork is a somewhat distant cousin of the heron and, in a general way, like it in structure and food habits, but a little less in stature and reach of wing. It does not exist at all on the western hemisphere, but is abundant through much of Europe, Asia, and Africa.

While the heron is a creature of the wilds and the solitudes undisturbed by man, this cousin of the Old World has, through the centuries, shown a liking for human association and has always made its home where man dwells.

In Holland, Denmark, and Germany the coming of the stork is proof that springtime is at hand. This big bird winters away down in the tropical swamps of Africa or of India, but comes north for the nesting. It arrives on the northern fringe of Africa and builds its nests, even as early as February. It chooses the low thatched roofs of the natives as its favorite nesting place, and a village of twenty-five huts is likely to be made the home of fifty pairs of storks. Proud is the native whose home attracts more than its average of two nests.

A month later the storks are nesting among the ruins of the ancient churches of Spain, adding a touch of life to many an old pile which suggests otherwise only things long dead. Still later the flocks have spread out in the low countries of western Europe and every nest has become the care and responsibility of some group of appreciative children. The flat tops of many chimneys form the foundation for such nests. On many a roof flat boxes are placed as a temptation to the visiting storks. A more ideal nesting place is furnished by many a Hollander who places an old cart wheel on a highpole as an invitation to the stork. Such a large flat base is just what they want for building and upon it they erect their brush nest, lined with some softer material, and to it they return year after year for raising their broods.

There are likely to be four or five young birds in the stork's nest, great clumsy fellows which are still quite helpless when they are two feet tall. The parent birds forage in the lowlands round about for frogs, snakes, mice, and insects, and feed these hungry mouths on the housetops. Soon the nest is so crowded that it is with difficulty that the brood can stay in it. It often happens that these ungainly youngsters tumble. They are always lovingly replaced, however, by their man neighbors, which fact suggests the thought that perhaps the stork nests on housetops so that it may get just this help in raising its young.

Still farther to the east in Europe, down in the Balkan peninsula, the stork is no less loved. No Turkish mosque is complete unless it has the nests of many storks upon it and no people regard these birds with greater reverence than do the Mohammedans. It used to be said in Turkey that no stork chose a Christian house top for its home. Greece, in those days, was ruled by Turkey. The stork came to be known in that section of Europe as the Turkish bird. It was therefore hated by the Greeks, and made its only man enemy of record through the centuries. When the Greeks had gained their independence from Turkey, they slaughtered great numbers of storks and tore down the nests. This dislike soon passed, however, and the stork has reëstablished itself even in the hearts of the Greeks.

Still farther east, in the Holy Land, in the valley of the Jordan, on the towers of the mosques of ancient Bagdad, is the stork at home. Huge flocks of them come up in the spring and spread out over vast areas. They seem to realize that by scattering out and allotting ten-acre tracts to each stork the foraging for each will be better, and the frog harvest will be more abundant. It is the hunting policy of the family, applied here by the stork as it is in far-away America by the great blue heron.

Storks fly surprisingly high in their migrations and are superbly graceful in flight, though somewhat ungainly when on the ground. Their white bodies, black wings, and bright red bills and legs give them a quite striking color scheme. They are odd in their manner of flight since they stretch their necks far out in front and allow their legs to trail in their wake—a habit which makes it possible to tell them at once from the herons, for those birds rest their heads between their shoulders while flying.

One of the ugliest creatures in all the world is the adjutant stork of Africa. It is a huge bird, five feet tall, with a beak a foot long and a shameless appetite. It likes village life also and pays its way by working as a scavenger. It often has been made a pet in the households of Englishmen living in Africa and has, upon occasion, proved itself very amusing. One such bird, for instance, was trained to stand solemnly behind its master's chair at mealtime and receive its share of the food. It was a great thief, however, and if not watched would gobble whatever it could steal. It was finally outlawed when, upon one occasion, it seized and swallowed a boiled fowl at a single gulp. Another such adjutant stork, which had the freedom of the household, was one day found in the act of swallowing the family kitten, but the master succeeded in catching the tail of that pet and pulling it back to safety.

QUESTIONS

1. The great blue heron heads the order of wading birds. What are some of its cousins?
2. Describe the place in which this bird lives. Tell how this solemn bird looks.
3. What does he eat? Describe his manner of catching water animals.
4. What odd developments of its body fits the heron into this life in the swamps? What strange facts have been discovered about his eyes? How do the eyes of hunting animals set in their heads? How do the eyes of hunted animals set?
5. Tell the story of a fight between a heron and a hawk.
6. Describe the flight of the heron when his solitude is invaded. Why does he keep to himself for hunting?
7. What sort of family life do herons live? What is a heron village like? Show that its site is wisely chosen. Where are the nests placed?
8. There is a night shift of herons. Describe the lives of these lesser waders. Where are they now to be found? Is the blue heron growing scarce? Why should pains be taken to preserve the blue heron?
9. What is the best-loved bird in Europe? Contrast the life of the stork with that of its American cousin. The white stork winters in the tropics. Tell of its coming north in the spring and of its nesting.
10. The Hollanders put a cart wheel on a pole as a nest. How many young birds are there in the stork's nest? Describe them. How do their parents feed them?
11. Why did the Greeks dislike the stork? Tell of their visits to the Holy Land. Describe these storks in flight.
12. What is the African giant of the wader family? Describe it as a pet. Tell the incident of the stolen dinner. Of the kitten.

CHAPTER XVI

THE OWL

SPECTRAL creatures of the night are the owls, that flit here and there on noiseless wings. Hollow voices they have, that thrust themselves into the lonesome solitude in which the traveler may find himself stranded, to inquire, as though from the grave: "Who? Who?" Or again, they may scream out in the night in high-pitched voices like human beings in agony. Unseen, ghostly spirits of the dark, the owls are individuals quite apart from other members of the animal kingdom.

Nobody needs to be shown how to tell owls from the other birds. They have no near relations. There are no other birds anything like them that might be confused with them. Through the ages they have had a large place in art and in story and their peculiar forms are imprinted on every mind. Upright they sit, looking wondrous wise. Perhaps this air of wisdom comes from the fact that they have such big heads. With all this room for brain why should they not be wise?

And those big, round eyes that blink so knowingly! Did you ever notice this peculiar thing about the eyes of an owl—the fact that they are set in the front of its head like those of a human being? The owl looks at you with its beak toward you and with both eyes. All other birds look at you from the side of their heads and with but one eye.

An owl actually has a face as has a human being. "The owl face" is quite different from the faces of other birds. It is a saucer-like face produced by ruffles of feathers that run around the big eyes and beak. The eye of the owl, like that of the cat, may expand its pupil more than may the eyes of most other animals. This lets in more of the small amount of light that exists at night and so the owl can see quite well at night. These pupils cannot contract to a pinpoint in the bright sun as can those of the day shift of birds and so the owl sees poorly by day. There is reason to believe that strong sunlight is quite painful to the eye of the owl.

It has a way, however, of protecting itself from the light. It has a shade, somewhat like those light-colored ones at our own windows, which it can pull across the eye when the light is too bright. This curtain is another eyelid, which all birds have, but which is probably more useful to the owl than to most birds. The eagle, also, uses it as a dimmer when he cocks his head and looks straight into the sun.

These eyes of the owl, looking to the front, cannot be moved very far from side to side. So it happens that when a burrowing owl is standing guard at its portal, and you walk around it, the bird slowly turns its head, following your movements. It has been said, not seriously of course, that if one walked round and round one of these owls, it would twist its head quite off.

Owls are deceptive birds. They are not nearly so big as they look. They are, in fact, always surprisingly small and skinny when one comes to examine them. They are little more than bundles of bones and feathers.

They have a way of fluffing out their feathers, however, that makes them appear much larger than they are.

In spite of size, however, the great horned owl, fiercest of them all, is no mean fighter. With one nip of his bill he can break the back of a squirrel or crush the skull of a rabbit.

A traveler in the Adirondacks was, one evening at sunset, passing through a clump of woods when he heard a great commotion going on in a near-by tree top. A moment later a big red-shouldered hawk, first-class fighting man of his clan, flew out, carrying a bleeding squirrel in his talons. It did not seem, however, as though catching a squirrel should have caused so great a stir in the tree top. While the traveler was wondering about it there appeared, in pursuit of the hawk, a great horned owl. These owls, flying in their quiet way, do not appear to have great speed on the wing, and a man is likely to be surprised when he sees them in action. One of them, for instance, is quite capable of capturing, while in the air, such birds as the swift-flying grouse.

This great horned owl pursued the red-shouldered hawk and, gaining a place just beneath him, darted suddenly up, seized the squirrel, and tore it from his rival's talons. Now it was the owl which was fleeing and the hawk that pursued. As the hawk swooped, the owl dropped the squirrel. Both birds then darted for it as it fell through the air, and both reached it at about the same time. They let it go and came to grips with beak and talons, making the feathers fly.

By this time the fighters had forgotten the squirrel and each was thinking only of how he might defeat his rival.

GREAT HORNED OWL STEALS RED-SHOULDERED HAWK'S DINNER

They circled about, each looking for an opportunity to strike. The owl seemed at first to miss his thrusts and it looked as though the hawk would be the victor. As they circled and clenched and circled again, however, the twilight was coming on rapidly. This fact began to tell in the owl's favor and against the hawk. The owl could see more plainly, while the hawk could see less well. The strokes of the owl became surer. The hawk became confused. In the end, torn and bleeding, he fled out of the woods with the owl, as victor, in close pursuit.

This great horned owl, nearly two feet long, is, as a matter of fact, one of the most ferocious creatures in the bird world, one of the hardest to tame. It has been known to attack a large tomcat and kill it. It frequently makes a meal off a skunk. One of these owls was once put into an inclosure with a great bald eagle. When morning came the eagle was dead. They have been known to visit turkey roosts and kill a number of these big birds in a single night, if not hungry eating only their heads and leaving their bodies yet fit for food. One owl, also, has been known to take two guinea hens off the roost in a single evening and devour them near by, leaving a circle of feathers in the snow as proof of the excellence of its appetite.

The great horned owl is the first of all the birds to establish its nest in the spring and to lay its eggs. It will find a hollow in a tree if possible. If not, it is likely to select the deserted nest of a hawk, or of a crow, and there to settle, raising a family as early as February, when the blizzards of winter are still raging.

The snowy owl, sentinel of the North, is even bigger

than the great horned owl, but a milder creature, feeding almost entirely on lemmings, mouse-like animals of the North, and on the big Arctic hares. It hunts by day, sitting statue-like on some high point, and pouncing when its game appears. In the South it eats mice and rats chiefly, thus performing a service which is very helpful to the farmer. This snowy owl lives all around the pole in the tundras of the far North and comes south only in the winter time. It is often quite abundant at this season in New England and in the other states near the Canadian border.

The barred owl, which is widely distributed, is a little smaller, and is without ear-tufts. It hides away in the deep forests, and is likely to hunt its prey in marshy tracts. All owls have very highly developed ears, and this barred owl, in particular, is believed to flit noiselessly about and listen for sounds made by any living creature below. It can hear the patter of the feet of the shrew on the snow, or even the stir of the mole as it runs its tunnel underground.

It is in summer that human beings are most likely to become conscious of the barred owl, through its melancholy hooting in the twilight. Two such birds station themselves in trees some distance from each other and call back and forth. In the South it is held that the first bird says: "Sal cooks for my folks, who cooks for you-all?"

To this the second bird replies, a little more faintly through a greater distance in the twilight quiet: "Sal cooks for my folks, who cooks for you-all?"

This barred owl, like the great horned owl, seldom stirs

by day. This may not be in either case because it cannot see well enough, for the great horned owl, at any rate, is by no means blinded by the light. There is another reason. The crows, the jays, and, in fact, the smaller birds in general, instinctively recognize the owls as their natural enemies. They seem to know and resent the fact that owls steal upon them while they are asleep and devour them. So, whenever an owl, even one that does not harm them at all, stirs by day, all the birds in the neighborhood are likely to set upon it and so torment it that it is driven to its retreat in the dense forest. It is partly to escape these pests that the owls hide through the day.

The screech owl is one of the best known of these prowlers of the night. It is an owl with a strangely quavering voice, which sometimes suggests the neighing of a horse, yet so melancholy and faint as to leave an impression that it is some ghostly charger of the past. The screech owl is a small bird, rarely ten inches long, given to making its nests in the deserted holes of woodpeckers and to retreating into hollow trees and chimneys. It may even happen that a housewife, opening the lid of her cooking stove of a cold morning, may find that a screech owl has come down her chimney and made itself comfortable within.

It is an odd thing that these screech owls are sometimes reddish brown and sometimes silver gray, and that both colors are often found in the same family.

All the owls have a great talent in their mottled coats for mimicking parts of the trees in which they live. Hearing the cry of the screech owl, one may look closely among the limbs, may see what appears to be a lichen-covered knot, and may be surprised to see that knot suddenly take wings and flit away. It is in fact a screech owl, thus making itself hard to see.

The fluffy feathers of these birds help them in fooling their enemies. One of them may be imitating a rounded lichen-covered knot on a tree and may fluff out its feathers for the purpose. Then again, it may be playing the part of a slender stub and may draw its feathers close about its thin body and appear a very different thing.

Another of those smaller owls that is well known, and a source of much interest to its observers, is the burrowing owl of the West which lives in the ground, its favorite home being the hole of the prairie dog. This burrowing owl is a handsome, harmless, and in every way lovable little creature, said, without any foundation in fact, to live in the same burrow with prairie dogs and rattlesnakes. It is much given to coming to the surface, standing guard quizzically at the door of its dugout, and bowing politely when you approach.

Of these small owls, however, the most important and, except in the northern part of the United States, the best known is probably the barn owl, or monkey-faced owl, which is given to finding itself a home about the buildings of man and, though little seen, is much heard as it makes high carnival by night.

A very striking fellow is this barn owl, a foot and a half long, with his white and tawny speckled breast, his gray and yellow back, his ruffle of feathers about his face that makes of it a big, white heart, hung up for all to see. He lives quite happily in the barn, never disturbing the swallow, sparrows, or pigeons that nest about him. He comes out in the evening, launching himself with a quite startling scream on his hunting in the twilight. After that he moves noiselessly on muffled wings, as is the

THE BARN OWL MOUSE CATCHER

way of his kind, wings that make no swish as they cut through the air. They make no noise because the feathers of them are fluffy instead of stiff and harsh. The owl, of all the birds, is fitted with wings that are to it what rubber heels are to men who walk on the ground. They are silencers of its movements. It goes on a business, here in the evening, that calls for stealth.

What is probably the most famous pair of barn owls in all the world dwelt for a long time in the tower of the Smithsonian Institution in Washington, quite surrounded by scientists who were taken up with investigation. Just across the street was the Biological Survey, the government bureau which studies birds to find out whether they help or hinder man in his attempts to raise crops to feed the multitude.

Whether birds are good or bad, from this practical standpoint, depends upon what they eat. If they eat the crops that man raises, and lessen the food supply, they injure him, but if they eat other creatures that prey upon crops, and increase the food supply, they help him. This food supply is a most important thing, so the government experts study, from all points of view, the problem of making it more abundant. Now, about the owl, they wondered, was it helping or hurting?

They wanted to find out what the owl was eating as a regular diet. They had to kill most birds and examine the contents of their stomachs to find out what they ate. These owls in the Smithsonian tower, however, furnished the information in a much simpler way than by dying to do it.

Owls are very greedy it seems. They can easily eat

their weight in food in a single day. They are not finicky about it. They gulp their food whole. Shells, hair, feathers, bones, all go down. If the thing they have caught is too big to swallow whole, they may tear it into two or three pieces, but they never pause to pick the bones.

They have a mill inside them that separates the good from the bad. That mill rolls the bad parts up in a pellet, puts it on the elevator, sends it up, and the owl discards it. There is no need of killing the owl to find what food is in its stomach, when, by examining these pellets, one can get the same information.

One of these scientists examined two hundred pellets thrown out by the Smithsonian owls. He found the skulls of two hundred and twenty-five meadow mice in them, of one hundred and seventy-nine house mice, of twenty rats, twenty shrews, six jumping mice, two pine mice, one mole, and one sparrow. By this and other similar studies was the fact established that the barn owl is the greatest mouse-eating bird in all the world.

All the owls have been studied in a similar way. It has been found that all of them are unceasing eaters of mice and rats. These are the chief food of all the breed. The little fellows, like the burrowing owl and the screech owl, eat insects as well, but they still eat rodents. The big fellows, like the barred owl and the great horned owl, eat birds and sometimes poultry, but they ceaselessly devour rats. They work by night when rodents are abroad, taking up the task where the hawks leave off. They are thought to be more effective even than the hawks, more effective, in fact, than any other enemies of these pests.

Insects and rodents are the farmer's worst enemies. The house rat and the field mouse are near the top of the list as destructive animals. They destroy uncounted thousands of tons of grain every year. The cotton rat in the South does great damage to the staple crops down there. The gopher in the West eats bark off the fruit trees and they die.

Everywhere the silently flitting owls are catching and devouring these rodents. One barn owl will do away with as many of them as a hundred house cats. Where the owls are abundant rodents become scarce. Where owls are scarce they become abundant. It is the owl, more than any other natural enemy, that keeps the rodents in check.

It is certain that the owl is among man's best friends in the bird world. The great horned owl possibly does him more harm, through killing poultry, than it does good, though this is doubtful. The barred owl kills an occasional half-grown chicken, but pays the bill many times over in rodents destroyed. The deeds of the others are nearly all pure gain.

Yet man, through the years, has regarded the owl as his enemy. Nearly all men everywhere still so regard him. Few men with guns on their shoulders fail to shoot him on sight. Great states, as, for instance, Pennsylvania, have placed bounties on heads of owls, have paid premiums to those who would kill these good friends of the farmer.

The public has never made a greater mistake than this in the treatment of any element that goes to make up its wild life. Government scientists have demonstrated that the owls should be protected and not destroyed, that their protection and increase will result in the decrease of the rodents and an increase in crops. Good citizens should become informed on this subject and see to it that owls are protected in the communities in which they live.

QUESTIONS

1. Why is there little danger of confusing the owl with other birds? What sort of feeling comes over you at the mention of owls?
2. What is there about the way the owl looks at one that is different from other birds? Describe "the owl face."
3. How is it that the owl can see better by night than by day? How does it protect its eyes from the bright light of noonday?
4. How do owls manage to appear bigger than they are? Give examples of the strength of their beaks.
5. Tell the story of the fight between a great horned owl and a red-shouldered hawk. What part did night and day eyes have in this fight?
6. Recount some of the exploits of the great horned owl. When and where does it nest. What owl is bigger than the great horned owl?
7. What do the barred owls say to each other? Have you ever heard them? Where are they to be found?
8. What is likely to happen to the owl that stirs by day? Why is this attack made?
9. Describe the night call of the screech owl. Where do they nest? What strange fact is known about their coloring? How do owls hide by making themselves difficult to see?
10. Describe the practices of the burrowing owl. Which is the most important of the smaller owls? Tell of his hunting in the twilight. Why do his wings make no noise?
11. How do scientists tell whether birds are helpful or not? How was this task made easy by a certain pair of barn owls? What do barn owls eat?
12. What do most owls eat? Is this helpful to man? How should man rate owls among his animal friends?
13. Despite its helpfulness to him, what has man done to the owl? What mistakes have communities made? What action, if any, has been taken toward the owl in your county or state? Is this right?

Chapter XVII

THE ENGLISH SPARROW

A BARBER in New York City once put an English sparrow egg into the nest of one of his canaries. It hatched and grew up in the family of this, its cousin. Strangely, and despite the fact that all its kind are given only to an unmusical chatter, it learned the canary songs which, by the way, it sang with sparrow energy. It became the pride of the barber shop and a marvel to visitors.

One day the door of the cage was left open and all the birds escaped. The canaries, having lived for generations in cages, allowed themselves to be caught and returned to their home. But the sparrow flew away and hustled for itself in the gutters of the great city.

Months later a little girl found a sparrow with a broken wing, carried it home, nursed it back to health. To the great surprise of the family it one day burst into song, a song like that of the canary, yet with a good deal of sparrow character in it.

A reporter heard of this strange singing sparrow, and wrote about it for his paper. The barber read the story, claimed the bird, and, despite the fact that all English sparrows look much alike, finally established his ownership and brought it back to his shop.

The English sparrow that had this peculiar experience

was a member of that foreign race, members of which are often accused of being the arch criminals of the bird world. All its kind are combative sort of persons, and, wherever found, are likely to be in conflict with man, beast, and their fellow birds. They bring in their wake discord, disturbance, and dirt. They have many shortcomings and but few virtues to make up for them. They are often referred to as the rats of the bird world.

Yet the English sparrow is an important bird that cannot be ignored. There are many respects in which it surpasses all other members of the feathered world. It is, for instance, better known to man everywhere than is any other bird. It plays about more doorsteps than all the others put together. In city, town, and hamlet, wherever the dwellings of men are clustered together, there is likely to be found the English sparrow. In the wilds where there are no men there are no sparrows of this species. So close does it stick to the abodes of man that, except in America, it is called the house sparrow.

There are more English sparrows than birds of any other breed. They are so evenly distributed throughout the United States, for instance, and are so abundant that they have no rival in abundance. They are equally plentiful throughout Europe, Asia, Australia, New Zealand, and Hawaii. They have come to blanket the temperate zones of the world.

The bird deserts of the world are the hearts of great cities. There are few birds where blank pavements and houses stretch on endlessly. Often, aside from the English sparrows, there are no birds at all. There are millions of children in Paris, London, and New York who, but for this sparrow, have never seen any bird outside a cage. It makes up the whole of the bird life of the tenement districts. Other birds may be creatures of the open spaces, of leafy trees and grassy glades, but the English sparrow belongs to the bustle of cities, is happiest where congestion is greatest.

There are many kinds of sparrows, of course, besides this house sparrow. They all look a great deal alike, just as terriers, collies, and hounds are all recognizable as dogs, but they have minor differences. They are mostly plain brown birds, but there are endless variations in the patterns and cuts of their coats.

The song sparrow, modest and gray, dearly loved throughout the United States, would be recognized anywhere by its coat as a relative of the English sparrow, yet how different is its nature! It can be identified, out there along the border of the woods, by a small, black spot in the middle of its breast, and by an odd way it has of "pumping" its tail as it flies.

The field sparrow, a shy, white-breasted, pink-billed bird, and a good singer, shows in its looks its kinship to the sparrow from Europe, but is quite unlike him in manners. So are the trim swamp sparrow, the gray and modest vesper sparrow, the speckled fox sparrow, the dainty and light-colored tree sparrow, and all the rest of them. Altogether, it is a very large and well-bred group of birds of which this one member has become a rowdy, greedy ruffian, given to mere noise and cluttering up the place.

Then, there are the first cousins to the sparrows, the finches, the linnets, the grosbeaks, the buntings. Truly a large bird family!

There might be a bit of interest in trying to see how these bird relationships work out, just how it is known that certain birds group with certain others. The kinship of the finches and sparrows, for instance, both big tribes, is shown by certain common peculiarities, upon which the scientists depend in their classifications. All these birds have, for one thing, the same number of big feathers, primaries, they are called, in their wings. They all have nine primaries. All of them have similarly shaped bills. They have short, stout, cone-shaped bills. All have well-developed gizzards.

These birds all have stout bills and strong gizzards for the same reason. They are all seed-eating birds. That is the outstanding characteristic of them all. They forage all the time, everywhere, for grain, weed seed, tree seed. They use their stout bills to shell these seeds, sometimes to crack them before swallowing them. So strong are their short bills that almost any of them can crack a grain of corn. Their gizzards must be built for grinding grain. Insect-eating birds have less need for such highly developed beaks and gizzards.

Scientists like to make orderly arrangement of whatever they are studying. They divide the animal kingdom, for instance, into "branches" such as the vertebrates, which have backbones; and the invertebrates, clams, for instance, which have none. Then they divide the vertebrate branches into "classes," such as birds, which have feathers; and the mammals, which feed their young on milk.

The first division into which the bird class is divided is called an "order." All the more familiar small birds, for instance, typically living in or about trees are grouped in the order of perching birds. Then all the seed eaters with their conical bills and nine big feathers in their wings are put together in the family of sparrow-like birds. The sparrow or finch family again is divided into "genera," the singular of which is "genus." Each genus is divided into "species," and we have the house sparrow genus, the song sparrow genus, and so on.

Setting this down in outline form we have a picture of the method of the scientist in bringing order out of chaos in this world of so many living things. It looks like this:

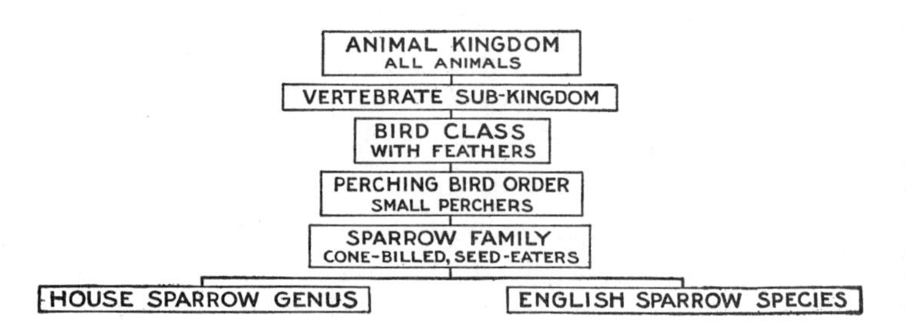

This finch family, including the sparrows, is the biggest bird family in all the world. It compares in birdland to the Smith family among English-speaking peoples. When one begins to look through the familiar birds for the peculiarities that would put them in this family, he finds that many old friends belong here. The canary, for instance, is a cousin to the sparrow, a cone-billed seed eater from the Canary Islands and the Azores, out in the Atlantic, and from Gibraltar.

The American goldfinches, or wild canaries, noisy and

happy rovers throughout the United States, belong to this family. The crossbills which use their oddly formed beaks for foraging the seeds out of pine cones, are members of this seed-eating group. The purple finches and the house finches, lovely colored, sweet singers of the East and West, are sparrow cousins. So are the snowflakes, or buntings, and the juncos, or snowbirds, that spend the summer in the far North and come down into the middle latitudes for the winter. So is the radiant cardinal, shy dweller in the thickets. These and innumerable others belong to this, the biggest family in birdland.

It would be hard to find a bird in all the world with a stranger story than that of the English sparrow; with a life history that has been more closely tied up with that of man.

The English sparrow is native in Europe. There, no doubt, it picked crumbs about the abode of man before the time he first began to put words on paper. Through all the ages of the development of the western peoples, there in Europe, there has never been a door yard that did not every day, summer or winter, know the presence of this sparrow. It has been the ever-present companion of every European through the passing centuries.

It is little wonder then, that when these Europeans came to America where there were no house sparrows and established new homes, they should miss this bird that had come down with them out of the past. During the early life of Europeans in America, Australia, New Zealand, and other distant lands they found not one of these birds.

From 1850 to 1870 in the United States the introducing

of English sparrows was a great fad. They were brought from Europe and sold at fancy prices to enthusiasts in many cities. Philadelphia, for example, deliberately brought over a thousand sparrows and turned them loose

THE AMERICAN CROSSBILL

in its parks. People in many parts of the country sent for small lots of sparrows and released them, thus starting colonies.

A typical instance was that of a gentleman in Topeka, Kansas, who, in 1864, imported twenty-eight of these birds.

He kept them in cages until all but five of them died. These five he turned loose, little expecting that they would survive. They gathered in a little group near by, however, and the following autumn there were twelve sparrows in the group. The second season there were sixty, the third season there were three thousand, and after that the numbers were beyond counting. This group did much to establish the English sparrow throughout the Middle West.

This bird was scattered in the first place by persons who believed that they would help the trees, and in the second place by immigrants to America who had grown up with these sparrows about them and had therefore formed a sort of affection for them.

After the sparrows had thus made their homes at certain points, they found ways of their own for getting widely scattered. Nothing delights an English sparrow more, for instance, than to hunt for food along the tracks at railway stations, and to examine freight cars on the side tracks. Grain may have been shipped in some of these cars and much of it may still be scattered about. The birds would hide beneath the eaves of such cars, and would even build their nests in them. Presently, a brakeman would come along, close the car door, and shunt this "empty" into a string that was bound for some distant city. The captive sparrows would ride across the continent. Then the door would be opened, and they would hop out and start a sparrow colony in a new community.

A single pair of sparrows would thus arrive in Denver. The first season it was there it would probably produce four broods of birds, with five little ones in each brood. Thus there would be two or three dozen sparrows by the second summer, and hundreds, multiplying into thousands, in the few years that followed.

By such methods as these did the English sparrow in America in fifty years blanket a continent. In such a way has it established itself in many lands all around the world.

Having thus gotten well started, the English sparrow proceeded to show its American friends that it was not a thing so much to be desired as it was a nuisance to be endured. In the first place, these birds, increasing rapidly in numbers, became very abundant in the cities. They nested in great numbers about public and private buildings and created quantities of filth. They were very uncleanly citizens.

As they became more numerous in the cities and towns they began mobbing the native species of birds to be found there—the song birds, the insect catchers, those birds, like the oriole, that were popular because they were beautiful. The English sparrow instead of a song had only a noisy and unmusical chatter. Instead of eating insects which would have served man's purpose, it chiefly ate grain which to him meant waste. Instead of being beautiful it had only a plain, brown, pudgy figure.

The English sparrows were gangsters, ruffians. They attacked the warblers, the thrushes, the orioles, the flickers, and drove them out. They seized the nesting places and, when other birds had already built their nests, they tore them down. Thus they put themselves in the places of more attractive birds that they had driven away.

These English sparrows were at first birds of the town

and cities, for they lived only around the abodes of man. As their numbers increased, however, they were crowded more and more into the country and began the same outrages around the houses and barns of the farms that they

THE ENGLISH SPARROW GAUSTER AND HOUSE WRENS

had committed in the cities. In addition to this, they consumed steadily great quantities of that grain which is the farmer's chief product.

Thus was it finally shown that when the American people, not knowing what they did, deliberately introduced this European sparrow into the United States, they committed one of the most foolish acts in all their history. In Europe this house sparrow had been branded as a nuisance and a menace for centuries. In America students of bird life knew that it was undesirable.

In 1850, however, the United States had no official agency to protect it against undesirable bird, animal, and insect immigrants as it has now. If it had had such an agency it might have been spared the English sparrow.

A few decades later there was created in its Department of Agriculture, a Biological Survey, which is a scientific bureau of the government, the purpose of which is to study animals and birds in their relation to man and to advise the government on such matters as keeping out those which promise to become nuisances and fighting those nuisances that are already established.

The Biological Survey has for a long time studied the English sparrow. It has tried it in that court which it maintains for such purposes. Whether a bird is good or bad, in the opinion of the court, depends largely on what it eats. If it eats the insects that injure crops, it is likely to be helpful. If it eats the crops themselves, it is likely to be harmful. The stomachs of birds on trial are examined to procure evidence. The examination of sparrow stomachs shows that their chief food is seeds and grain, although they feed many caterpillars to their young.

Their food habits are chiefly harmful. This court also pronounces the English sparrow a rowdy, a ruffian, and a marauder. In many instances it places the sentence of death on it and advises citizens, where the bird is numerous and destructive, to tear down its nests, to trap it, to shoot it, to poison it, to resort to every possible means of reducing its numbers. This government bureau, as a public service, has printed many leaflets and bulletins showing the best methods of destroying English sparrows. It will send these bulletins and other information to any citizen who writes to it in Washington, asking for information.

The resentment against the English sparrow ran high a quarter of a century ago. Since that time, however, it has been growing less intense. Perhaps this foreign bird has adjusted itself to American conditions, learned to live more in harmony with other feathered citizens. Certainly it does not seem to attack other birds as it was reported to have done at an earlier time.

There occasionally comes into the vicinity of Chicago or Boston that terror of the north, the great American shrike or butcher bird. It slashes about, taking English sparrows where it finds them. Then it impales them on some thorn, breaks open their heads, eats their brains. The European starling is another introduced bird that meets the sparrow on its own heath and conquers it. The starling is fond of sparrow eggs and sparrow young ones and helps itself from the sparrow nests. Then there is the western house finch, which likes to nest in man-built bird houses. It comes into the stronghold of the English sparrows, fights them for their nesting places, and wins.

The English sparrow does not seem to be so boisterous a bully as it used to be. Maybe it has been tamed a bit by these rivals.

Of late the English sparrows are decreasing in numbers in the United States in sections in which they used to be most plentiful. This decrease is said to be due to the introduction of the automobile and the decline of the horse. The grain wasted in feeding horses used to be the mainstay of the English sparrow. Now that the horse has largely passed, particularly in the towns and cities, sparrows are less abundant.

There are still communities, however, in which there are too many sparrows. Utah, in the far West, is a state which has a very large English sparrow population. This is thought to be due to two or three peculiar conditions prevailing there. In the first place the farmers build sheds which are thatched with straw for their cows. These coverings for cow sheds make most attractive nesting places for sparrows and every one of them is well tenanted.

In Utah, also, Lombardy poplar trees have been planted along the irrigation ditches and the roadsides. There are grain fields everywhere. The English sparrows out there have done a very unusual thing for the breed. They have deserted the towns and the barn-yards to take to these poplar trees as nesting places and to the fields for food. So have they made for themselves a new manner of life.

They injure the grain crops in Utah. It has been found necessary to put on special campaigns to reduce their numbers. Yet in one time of need these grain thieves came to the rescue of the Utah farmers. A weevil had attacked the alfalfa crop which is very important in that state and threatened its destruction. The sparrows came to the rescue and ate up such quantities of the weevils that they overcame the danger.

QUESTIONS

1. Tell the story of the English sparrow that learned to sing the canary song.
2. What can you say about the nature of the English sparrow? How well known is it, and why? How do its numbers compare with those of other birds? What can you tell about its life in great cities?
3. What are some of the best-known cousins of the English sparrow? How do scientists tell that birds are related to each other?
4. Why do all the sparrow cousins have stout bills and gizzards? What is a "branch" of the animal kingdom? A class? An order? What are the divisions from "kingdom" to "species" in the animal world?
5. What is the largest bird family? What do you know about the goldfinch? Crossbill? Junco and the cardinals?
6. How is the house sparrow regarded in Europe? How were they brought to the United States? How were they spread? How did they steal rides on trains? Describe the growth of the little group at Topeka. How long did it take the English sparrow to spread over America?
7. In what way did these birds become a nuisance? How did they interfere with other birds? How did they spread to the country?
8. The United States made a great mistake when it allowed the English sparrow to enter its borders. What arrangement did it later make to keep this mistake from being repeated? If you now wanted to bring a bird from Europe to the United States from what government bureau would you ask permission?
9. When the government's bird court tried the English sparrow what was the verdict? Where can information be secured on how to fight this pest?
10. What swashbuckling enemy of the English sparrow comes out of the North? How does it attack?
11. Why is the number of English sparrows now growing less in the cities? What odd thing is it doing in Utah?

CHAPTER XVIII

THE BUTCHER BIRD

THE members of a certain country club had been much worried because the moles and gophers were running their tunnels beneath their golf course, and thus destroying the firmness and smoothness of its surface. Then one morning, they made a strange discovery. There, upon the course, was a gopher with its head neatly snipped off.

At first they thought that some player had used this gopher's head as a golf ball. A few days later, however, they found another gopher and a mole, each with its head gone. This looked curious and in fact developed into the club mystery. Then, one morning, an early player witnessed a combat which cleared up the whole situation.

A gopher had been unwise, as gophers sometimes are, had come from under ground and was stirring in the grass. An ashy gray sentinel, with marvelous eyesight, on a tree top near by, caught the movement and was upon the big rodent. These gophers, stocky built and the size of a house rat, are no mean fighters. When assailed they rear themselves upon their haunches and strike viciously with their big teeth. The ash-gray bird, itself not so big as the gopher, circled about, however, and watched for an opportunity to strike. Its strong, hawk-like beak went home at the back of the neck, just at the base of the gopher's skull, and the fight was over.

It was a late January afternoon in New Jersey, with a deep snow upon the ground and the thermometer playing around zero. An observer went into his garden and found, sitting upon a bush with his feathers fluffed out as is his way in cold weather, another of these ash-gray birds. This one did not fly until the owner of the garden was within four or five feet of him. He went a few yards to another bush and, upon the approach of the man, flew across the garden to a shrub in its corner. As he made this latter flight, however, a field mouse was crossing from the tool house to the garage, and the gray bird snapped it up.

The man watched to see how this bird would go about eating his game and found that the manner of it was quite odd. He selected a limb which forked, making a sharp crotch. He fitted the mouse into this crotch with the small of its back deep down in it, and its hips keeping it from pulling through. Thus firmly fastened, the bird began to pull the skin from its victim and then to tear it apart and devour it.

When he had eaten up to the small of the back, however, the hind parts of the mouse were left and the earlier scheme of fastening it did not serve the purpose. The bird then adopted another method. He took these hind legs of the mouse, selected a sturdy thorn, and spiked them upon it. As a later examination showed, the thorn passed through one of the holes in the hip bones. The bird showed that it knew a good deal about the make-up of mice by finding that hole with its thorn while the flesh was still on. With the mouse thus spiked it proceeded to finish its dinner.

A clump of thorn apple bushes skirted the road and back of them a brook ran down through a fringe of big trees. The whole was in a piece of land on the edge of a city where hunting was not allowed, and so the birds gathered there in great numbers. This was, in fact, a sylvan bird sanctuary.

Of a sudden a great clamor broke out among the clumsy grackles and they scattered and flew in all directions. Then the warblers began to chirp in distress, the song sparrows scurried about, and the orioles chattered noisily.

A careful observer looking for the cause would find that another of these ash-gray birds, white underneath, and with black wings, with white feathers in them, had come into this quiet bird paradise. Its appearance had created a great commotion and it seemed everything was at sixes and sevens.

Well it might be. If, on the following day, you had gone to the thorn apple trees you would have witnessed a gruesome sight. There, in a sharp crotch, you would have seen the limp form of a catbird, as big as the gray bird itself, first cousin to the mockers. You would have seen a little song sparrow stark and dead, hung there like a piece of meat at the butchers. On a thorn near by you would have seen a lizard impaled, while the remains of a little garter snake, half devoured, hung from another thorn.

This was truly a butcher shop of the forest. This bird which established it is truly the butcher bird, and is everywhere known as such. It is the terror of birdland and mouseland, a bold murderer and despoiler, a heartless and cruel sort of person that seizes upon whatever of small life comes its way and devours it.

This butcher of the bird world is quite an unusual citizen in feathers. It is a terror with which there is none to compare outside of those hawks and owls that are set down as birds of prey. The butcher bird, or shrike as it is also called, is not, however, a member of the family of the birds of prey. It is, in fact, more closely related to the sparrow group. The English sparrow, you will remember, is a rough, greedy bird, somewhat given to bullying. The shrike seems to have been, ages ago, a big member of the sparrow group that got started as a huntsman and kept getting rougher and stronger. It is put in that broader classification of sparrow-like birds, is a perching bird, by kinship even a singing bird, though its harsh life seems largely to have robbed it of its song. Its nearest relations

MYRTLE WARBLER KILLED BY SHRIKE

seem to be such birds as the top-knotted cedar waxwing and the dashing scarlet tanager.

There are many kinds of shrikes in the Old World, but only two important varieties in America. These are the great northern shrike and the southern or loggerhead shrike. These birds are much alike in form and color, the northern being a bit darker beneath. The northern shrike is about ten inches long, while the loggerhead is about nine inches long. Thus the latter is a little smaller and the former a little larger than the robin.

These shrikes, or butcher birds, as is the way of birds, seem to have come to an excellent agreement as to the range each should cover. The northern shrike nests in the north of Canada in the summer time, a region into which the southern shrike never intrudes. The southern shrike winters in southern United States and its cousin never comes so far south. There is a common ground, however, which one occupies in summer and the other in winter. While the big fellow is in the North, its southern cousin has moved into upper United States and lower Canada, a region which is the winter home of the Arctic butcher. Then, when summer comes again, each group shifts a thousand or two thousand miles to the south, and the gentleman from the North occupies the upper half of the States. Thus these two belts of butcher birds cover most of North America, shifting back and forth, without ever coming in contact.

The butcher bird has a big and business-like head, the head and beak of a hawk. Its feet, however, are the feet of a sparrow. It has nothing of the talons of the birds of prey. It does not use its feet in catching its prey, nor in holding it while it tears it to pieces. It is because its feet are not talons that it is forced to use crotches and thorns for anchoring its catch while eating it.

MIGRANT SHRIKE AND GRASSHOPPER

Here is a quite unusual thing about this butcher bird. It stores food. There are few birds that do this. Some of the woodpeckers drill holes in posts and put away acorns, and the blue jay hides large seeds and soft-shelled nuts, but there are almost no others. The butcher bird brings its catch to its shop, hangs it up to mellow, and keeps on hunting. This is particularly helpful in the winter time when meats keep well and when storms are apt to make the hunting bad.

One is likely to conclude, when he first makes the acquaintance of the butcher bird, that it is a cruel murderer and to condemn it forthwith. This is particularly true when one finds its shop hung with warblers and song sparrows. A careful study of its activities, however, makes it appear in a little less villainous rôle.

The Biological Survey of the Department of Agriculture has applied the test of stomach examination to the butcher bird. The result is, roughly, that they have found that these birds eat three kinds of food—insects, rodents, and other birds. Their food is made up of about equal parts of these three. The insects they eat are mostly big fellows, like grasshoppers and beetles. Eating them is serviceable to man. Devouring rodents, also, is a great help to man. In the South the shrike is often called the mouse hawk. It knows well when corn-husking time comes. When the farmer goes into the fields, tears down the shocks, and gathers the ears of corn, the butcher bird is very likely to take an active part in the work. It knows that there are likely to be a number of mice in each shock, and that they will scurry forth as the work advances. It takes its place on a near-by fence and watches closely. When the mice run for other cover the bird catches them. Doing so is a service to the farmers, for these mice are great grain eaters.

It was a rather unusually good turn that the butcher bird did when it killed the gophers on the golf course. Its method also was a bit unusual. These gophers and moles were, in fact, too heavy a load for it to carry. Also they were heavy enough to lie somewhat firmly when they were being torn to pieces. Under the circumstances the bird ate only its favorite morsel, the brain, of which it is very fond, and let the body go.

In the matter of bird slaughter, there is a great deal of loss in song birds because of the activities of the butchers. There is one fact in this connection, however, that must be set down to their credit. They do away with many English sparrows which are so plentiful that they have become a nuisance.

It is the northern shrike particularly that is the terror of the English sparrow. In the winter this bird comes into northern United States. The song birds have mostly gone south, but this noisy sparrow is still about the eaves and barn-yards in great numbers. The shrike breakfasts, lunches, and dines off sparrows. If they are plentiful it eats only the brain. Thus it becomes one influence, and there are few, that tends to hold these foreign sparrows in check.

So, weighing and balancing, back and forth, the Biological Survey reaches the conclusion that the shrike or butcher bird, north or south, helps man more than it harms him.

It is in November that both varieties of shrikes shift to the south. In March again they make the reverse move. On both these occasions they move as individuals. Most birds gather into flocks after the nesting seasons and migrate in great masses. This independent butcher, however, keeps to himself. No gathering in crowds for him. He is a solitary huntsman. He plays a lone hand.

In the spring the shrikes make their rough nests, usually in thorny bushes or low trees, and raise from three to six young ones. Unhappy indeed are the other birds in the neighborhood where the shrike makes its nest. A neighbor with the habit of butchery is very upsetting. The danger of death hangs over the little birds round about.

The destroyer, perched on a telegraph wire or a church steeple, looks out over his realm, occasionally uttering a harsh cry. Fortunately for the little birds this is insect time and grasshoppers furnish most of the shrike's food.

The watcher sees a prime insect go clacking its wings above the grass in its attempts to impress its lady love. He swoops and the grasshopper is his. See how low he flies toward the thorn bush—low with steadily flapping wings. It is a peculiar flight. Then he shoots suddenly upward into a tree where he knows there are suitable spikes. Here he impales his grasshopper, as is his way, and tears it to pieces. Such is the life of a butcher bird.

The butcher bird has a stout bill which it uses to tear the flesh of its victims as do the hawks. Some of its sparrow cousins can crack a grain of corn with theirs. A basic marvel is back of the ability of the bird to perform such tasks with its beak. It has given up its hands that it might devote them to its difficult specialty of flying. When it did this it was as though it had agreed that its hands should be tied behind it forever and that it would try to develop some other means for doing their work. It set about training its mouth to act as hands.

Yet this beak is but an instrument like a pair of pliers made of two stiff pieces of horn. It is remarkable that an oriole so equipped should be able to perform so effective a feat at weaving as it does in fabricating its hanging home. Yet this is but one of a multitude of tasks that birds perform with their beaks.

Their primary purpose, of course, is to secure the food which keeps their owners alive. The crow can dig into the ground with its beak for the grain of corn the farmer has planted or can crack an acorn for its kernel. The gannet which dives into the water and catches fish, admittedly squirmy and slippery, has barbs on its back teeth to keep its grip from slipping.

The pelican has a bag like a butterfly net beneath its chin in which it catches its prey. If the fish happens to be pointed the wrong way it must flip it into the air and change ends so it will slip down its throat. Having flipped the fish this bird must then catch it. To do this it spreads the sides of its under jaw which supports its net to many times its original width.

Many shore birds attempt to snatch the oyster from between its partly opened shells and sometimes they are caught by its quickly closing on their beaks and are held tight until the rising tide comes in to drown them. One of these waders, however, called the oyster-catcher, has developed a beak so strong that it can play the rôle of the oyster knife. It invites the bivalves to close on its beak. Then it tears them loose from their moorings, pries open the shells at its leisure, and devours the oyster within.

The orioles, as insect eaters, have stout bills that crush the life out of their victims, but narrower mouths than the chimney swifts and the night hawks that must scoop up their prey while on the wing, and less strength in their bills than the finches which crack surprisingly hard seeds before swallowing them.

The bill of the woodpecker is a chisel for woodwork and is an excellent instrument, but in New Zealand there is to be found the huia-bird with this business of grub hunting much more highly developed. The male bird goes about with his stout bill drilling into rotten logs and the female follows with a beak of different construction, a long curved probe with which she explores the wormwood below. In this way the family group boasts two instruments instead of one.

Aside from these specialized beaks there are some tasks that are common to them all. All birds, for instance, make their toilets with their beaks and, birds being dainty and careful as to their appearance, this is no mean task. In the animal kingdom only those cousins, the ants, bees, and wasps, devote more time than do the birds to keeping themselves well groomed. These insects have six handy feet with which to work, while the bird has but this one tool of all uses, its beak. With it the bird must earn its living, build its house, battle its enemies, sing its songs, and make its toilet.

QUESTIONS

1. What bird of the sparrow group can kill a gopher? Relate the incident of the golf course. Tell of another exploit of this hardy bird when the weather was down to zero.
2. How does the butcher bird eat its catch? Explain the use of a crotch; of a thorn.

3. Describe the coming of the butcher bird to the bird sanctuary. The fresh meat it hung up to keep. What creatures could be found there?
4. This butcher of the bird world is not a member of the family of birds of prey. What are its nearest relatives? Show how, as a huntsman, it has grown to be unlike the sparrow from which it came.
5. Another name for butcher bird is shrike. Name the two American kinds of shrikes. Where does each live?
6. The butcher bird has a beak like a hawk. How do its feet differ from a hawk's feet? What other birds store food?
7. The bird court has tried the shrike. Describe its work in the cornfield. Its attack on the English sparrow. Its slaughter of song birds. What conclusions does the court reach?
8. Describe the shifting of the two kinds of shrikes from the South to the North and back.
9. Describe the way in which this bird hunts. His peculiar flight. When does he eat grasshoppers?
10. What do the birds use as a substitute for hands? What became of their hands?
11. Relate some of the feats that various birds perform with their beaks. The pelican. The oyster catcher. The oriole.
12. What marvelous examples of two birds with different sorts of tools that practice teamwork comes from New Zealand?
13. Have you ever seen birds working with their beaks? How many tasks does a chicken perform with its beak? How do birds make their toilets? What is their chief weapon of defense?

CHAPTER XIX

THE BOBWHITE

A FARMER rode on horseback down a dusty road one summer day, his setter dog romping on ahead. Suddenly the dog came to point just beyond the barbed wire fence, stood like a statue, tail out, one front foot raised. Its nose, as the master knew, was directed at some game bird.

The farmer rode up to the fence and looked along the line of the dog's nose. It was leveled at the gray and brown form of a mother bobwhite sitting on her nest in the grass. Her colors so blended into the background of grass and ground that he could hardly make her out. She sat so still that there was no movement to catch the eye.

Then, as he watched, he saw a very peculiar thing taking place. There stole out from beneath the brooding bird a tiny brown object not much bigger than a cockroach. It went scurrying away into the grass. Then came another and another of those small creatures, each running in a different direction. Presently the mother bobwhite sprang from her nest and went whirring away.

The nest was quite empty. The farmer, wise in such matters, knew just what had happened. These little scurrying objects he had seen were tiny bobwhites, not a day out of the egg. They were hardly dry from the hatching and had never before left the nest. When this steady-eyed dog and this towering man animal appeared, however, the mother felt sure they brought the danger of death to her little ones. She softly sounded the signal of her kind which means, "Hide." Young as they were, her children knew what to do. Each slipped away, crept under leaves, and hid in the grass that so exactly matched their own colors. The mother waited until all had gone. Then she flew up, expecting to draw attention to herself, leaving but an empty nest behind.

It would have taken a most careful and minute search to have found one of these tiny bobwhites. The unskilled invader would not have known that they were there. He would have gone his way. When the danger was past the mother quail would return and would quietly call her brood out of hiding.

Such is but one of the emergencies that these plentiful game birds have to meet, and one of the schemes they employ to keep themselves alive. They have a hard time of it and the death rate is high, but their families are so large that, except when hunted too much by man, they seem able to keep up the numbers.

This quail is such an attractive bird from the standpoint of food that many creatures prey upon it. "Quail," did I say? That is its name in the North and West. Perhaps you live in the South and are used to hearing it called "partridge." Perhaps, again, you live in some section where it is known as the "bobwhite," as it is generally through the Middle States. Bobwhite is a compromise name along the borderline where people disagree as to whether it is a quail or a partridge. Bobwhite is its own name for itself, and where could there be a better authority? There is a better chance of avoiding disagreements in calling it bobwhite rather than any other name.

Although known by all these names, it is the same bird everywhere in America east of the Rockies. West of those big mountains there are half a dozen varieties of quails, or partridges, very much like those of the East in their manner of life, but quite different in feathers and coloring. The bobwhites of the East are mottled gray and brown and the males and females are much alike. The quails of the Pacific Coast are much darker in their coloring, the males often being quite radiant and boasting very attractive topknots. The eastern bobwhite lays white eggs, while its cousin in the West produces a very speckled output. The western quail, also, does not say "bobwhite," but has a call all its own. It, therefore, could not be called a bobwhite. Otherwise, however, the habits of the two varieties are much the same.

The quails, bobwhites, partridges belong to the scratching birds. There are comparatively few of these scratchers, of which grouse, chickens, and turkeys are the outstanding representatives. They are of those birds that are called "fowls." They are ground birds, very accomplished as runners, but poor at flying. They fly only short distances on stubby wings. They scratch the ground to uncover food. They eat grain and weed seed largely, though they are likewise fond of insects. The bobwhite is the smallest of the American scratching birds, yet is is the most generally prized of the game birds.

The bobwhite writes the story of the danger that it faces when it makes its nest so snugly there on the ground among the grasses. It lays in that nest from ten to twenty eggs, more than almost any other bird. They are pointed eggs that fit like paving stones into this cup-like nest. They would not fit thus if they were not so pointed.

Since it is an established fact that the birds that face the most dangers lay the greatest numbers of eggs, it

BOBWHITE AND HER YOUNG ONES

would seem that the bobwhites are poor insurance risks. They are indeed poor risks, because, in the first place, they are so good to eat that all manner of creatures hunt them and, in the second place, they live on the ground and have no great power of flight, and are, therefore, comparatively easy to catch.

The native wild cats, and that vast crew of house cats that have taken to the brush, all prey upon the quails. The prying fox finds them and takes his toll. The creeping snake steals upon them, there on the ground, swiftly strikes, lathers its victim to make it slippery, and swallows it whole. The lumbering skunk, the treacherous weasel, and the soft-flying hoot owl, make night dangerous for them. By day the shadow of the sharp-shinned hawk flits over the field, its owner darts like an arrow, and often drives home its tong-like talons before the foragers can reach cover. Then, finally, there appears the man creature himself, with a dog and a gun, depending upon this dweller in the stubble for his choicest autumn shooting.

The bobwhite hatches a dozen or fifteen young ones and is lucky if three or four of them get through to the breeding season of the following spring.

Yet this bobwhite family is a happy group and one highly prized by man, with whom it dwells. In the early days, when America was sparsely settled, a traveler in the great wastes could always tell when he neared a settlement by the increase in the number of quail. The weedy fields and waste from the harvest give the bobwhites unlimited food and lead to their increase. Man with his gun takes great toll from their numbers and thus tends to keep them down. So is the balance struck.

As America has been settled, however, and man has become more and more numerous, it has been necessary to regulate the bobwhite shooting that they might not all be killed. Now they are protected almost everywhere, with generally an open season in the autumn which, by careful management, may be so adjusted as to keep the birds at the desired numbers.

Thus it comes to pass that the bobwhite family lives in pleasant relation to its man neighbors for the greater part of the year and needs match its wits against his only during the brief shooting season. It feeds in the stubble after harvest, picking up grain that would otherwise go to waste. It eats almost no grain before harvest. It renders yeoman service to the farmer, saving him much work with the hoe, by eating weed seed. Lucky, also, is he to whose garden a flock of bobwhites comes to feed, for they are insect pickers of a high order.

These baby bobwhites that are so active, even upon the day of their birth, and scurry to cover when danger presents, are in this respect quite different from most of the youngsters of the feathered world that are so helpless in their nests, and must be pampered, fed, and brooded for days and weeks before they are to any degree able to take care of themselves.

The young of the scratching birds go with their parents to the food rather than having the food brought to them. No worm carrying and bill feeding is needed for them. No sooner is the bobwhite brood hatched than it is taken on expeditions of exploration about the world in which it lives. Wandering through the brush or the stubble, the parent bobwhites find grain and weed seeds and insects to which they call their young, much as the barnyard hen feeds her chickens.

The farmer boy coming barefoot down the cultivated corn rows may burst in upon a flock of bobwhites in the privacy of its dust bath, fluffing the dry dirt into its feathers as do chickens in the ash pile. This farmer boy may surprise the group and leap into it with the idea of capturing some of the young birds. At the alarm cry of the mother they scatter instantly, while she flings herself fluttering beneath the very feet of this rough intruder. She seems almost within his grasp and he is likely to turn his attention to her and forget the little ones. He almost, but not quite, catches her, and she flutters a little way, leading him off, giving her young a chance to hide, and then, quickly recovering, goes flying away.

These young bobwhites are very much more active for their age than are even young chickens and learn to fly before they are half the size of their parents. The family is now a group of free-roving foragers of the summer, leading a happy life of unrestraint in the stubble and ragweed. It goes at night to roost on some hillock which gives it a point of vantage in watching for the creatures that prey. Wise is the bobwhite family in going to this roost, for it always flies for a distance before reaching its night's lodgings. It flies because the air leaves no trail that might be followed by the sharp-nosed beasts of prey.

At the roosting place the group arranges itself with all the skill of a trained army troop in the enemy country. It forms a circle with the tails pointing in and the beaks pointing outward. In this formation there are eyes looking in all directions. In this formation also, in case of surprise, every individual can plunge instantly straight ahead and into the air without interfering with its fellows. In case of trouble, the group can and does explode instantly as might a bomb.

Then come the autumn days when man goes into the

stubble with dog and gun. It is a good place to hide, especially for birds that can run swiftly among the ragweed and the still standing grain stalks. Usually the bobwhites select fields with woods or briar patches adjoining, and these are their refuge in case of trouble. Man, single-handed, even with his gun, would not be so very dangerous to the birds. He has, however, trained his dog to help him, a sharp-scented dog, with a marvelous instinct for finding them out and a knack of pointing to them for his master's information. The way the two work together is very interesting.

The bobwhite covey may be hunting in the stubble. The birds hear the beat of the feet of the galloping dog, the mother sounds the danger call, and the youngsters hide in the weeds near her. They stand very still, fading into the background, depending on their color camouflage to make them invisible. The father sits at the edge of the covey, ready to lead a dash for cover if it becomes necessary. Then the dog's head appears above the stubble. The birds sit very still. They are ready for flight, but, until the enemy approaches more closely, it seems wiser to depend on not being seen. They and the dog remain motionless like a posed tableau.

Then there comes the tramp of the man. This man with a gun is a rather new thing in their world. They have certain instincts for avoiding the enemies that have always been about them, enemies with whom they have been contending for hundreds of thousands of years, but this new foe is a puzzle, a terror to them. For most of the year he seems their friend. Then of a sudden he becomes a monster belching thunder, flame, missiles of death. They have not yet developed instincts with which to cope with him.

If the man is an amateur sportsman with a half-trained dog, they are safe enough. But these wonder dogs, followed by deadly shooting hunters! They are different. The bobwhites sit still until the approach of man proves to them that the waiting game will not do. Then, of a sudden, the covey explodes. Led by the father, the birds spring into the air on whirring wings. It is wonderful what speed they can develop in a few yards. They spread out fanlike and strike for the cover of the woods.

It is then that the gun of the man belches fire and death. Bang! Bang! And with each explosion a member of the covey changes from the live, taut, pulsing, vigorous thing that it is, to a slack, lifeless mass that goes tumbling through the air, then drops into the grass while feathers trail slowly away in the wind.

Nor is this the end. This man with his dog beats up the birds that have not joined the flight and they are picked off one at a time. Then, perhaps, he follows the frightened flock into the woods, flushes one after another and brings it down.

Presently, when quiet has come, the mother may be heard sounding her assembly call. "Ka-loi-lee! Ka-loi-lee!" she shrills. To this there comes the answer here and there, "Whoil-kee! Whoil-kee!" and the covey, what is left of it, is again brought together.

It is hard on the bobwhites, this month of shooting, yet not a matter over which one should grow too sentimental. Man's stubble fields have brought to them such abundance as they have never known before, have created conditions under which they would breed out of all proportion but for this restraining influence. They are dependent on man, are almost domesticated birds. They are game birds, sources of sport and of food. They may as properly be killed as are chickens and cattle. Their killing should not be entirely stopped. That would quite upset things. They would soon become too numerous, and might become pests. Shooting should be and is being controlled, allowed in the proper degree to maintain the balance.

Bobwhites are being preserved in comparative abundance. Almost everywhere now their fascinating whistle is heard in the spring wheat fields. "Bob, bobwhite," they call, "Bob, bobwhite." It is the mating season and this call is the challenge of one of the cocks to the other. Two rivals sit on widely separated fence posts and whistle back and forth to one another. In the country there are many interpretations of these calls. There are those who say that they call first one and then the other, and this is what they say:

"Bob, bobwhite!"
"Are your peas ripe?"
"No. Not quite."
"Will your little dog bite?"
"Yes, sometimes
In the night."

It seems, however, that the talk is of something nearer to their hearts than this, because the outcome is often battle. The issue, in all probability, is Miss Speckles, sitting demurely there in the grass, and it is not unlikely that the victor wins her as his mate, his partner in that long fight that is to follow for the successful bringing up of a family.

BOBWHITE SINGING FROM A STUMP

The bobwhites, being but short distance fliers, migrate little, or not at all. Being seed eaters they forage through the winter in the grain fields, among the ragweed, the fox-tail grass, even about the barn-yard. There is usually open ground somewhere round about where seeds may be picked up, or tassels sticking above the snow, from which they may be gathered. Times of heavy snow are, however, danger times for the bob-whites. There is danger from starvation, from freezing, and strange and unusual dangers in roosting places under the snow.

Even in winter the bobwhites continue to

be ground dwellers. When a heavy snow comes they are likely to take refuge in some such place as a fence corner where a fallen bough has caught the drift and built beneath it a white arched temple which might be a fit home for a fairy. The doors may have been wide when they went into this home under the snow, but the white mantle may have kept falling through the night and may have blocked up the entrance. Unless it is too deep, however, this need give them no worry for they know how to break their way through the fluffy snow.

Sometimes, however, conditions conspire to bring tragedy to the bobwhites. Maybe toward evening, when the flock is tucked away, the weather grows warm and the snow tends to melt. Then it turns chill again and this melting snow freezes into a glaze of ice. When morning comes the birds find themselves as snugly locked in as though a metal tank enclosed them. It may happen that the trunk of a bush, running through the snow, has blown back and forth and kept a passage open. It may happen, on the contrary, that they are sealed into a small chamber so tightly that not a bit of air can get in. They may be smothered as they sleep there in their family circle, bills out. They may be held there overlong and starved. It is no uncommon thing that, when the snows melt in the spring, family groups that have thus perished are revealed.

Yet good fortune may intervene and snatch the prisoners from the grip of tragedy. A stone thrown at a woodpecker by a boy on his way to school, an icicle shaken from an overhanging tree, the crunching foot of a prowling donkey, may break the crust and bring a reprieve before the prisoners die. It may even happen that knowing man neighbors may give thought to the bob-white flocks, may search them out, and may release them and provide food until the snow is gone.

QUESTIONS

1. Creatures of the wild, in the face of danger, are guided by an instinct that often saves their lives? Show how the bobwhite mother hid her brood when she thought it in danger. Show the wisdom of the newly hatched birds.
2. What different names are bobwhites called in different sections of the country? How do the birds of the East differ from those west of the Rockies? How do their eggs differ?
3. Bobwhites belong to the order of scratching birds. Name some of their best-known cousins. How well do the scratching birds fly? What is their principal food?
4. Why are the bobwhite's eggs so pointed? What does the number of them show as to the danger of death which this bird faces? What are its chief enemies?
5. Bobwhites like to live near farm lands. Why? What good do they do?
6. How are the feeding habits of young scratching birds different from the habits of other birds? Describe the life of the bobwhite family.
7. Where does the bobwhite family roost? How does it fool beasts of prey? How do the birds group themselves for the night? What happens if they are disturbed?
8. Man, with his dog, goes bobwhite hunting in the autumn. What do the birds do at their approach? Describe the flight. The death.
9. Is it right that man should shoot bobwhites? How are they being kept at the proper balance?
10. To what extent do bobwhites migrate? What do they eat in the winter time?
11. Describe the bobwhite roost under the snow. How may they be trapped and smothered? How, on the contrary, may they be saved?
12. What do you think of the life story of the bobwhite? What birds, if any, do you think are more interesting and why?

Chapter XX

THE WOODPECKER

HERE is a block of wood, a section of the body of a young tree three inches thick. It has been split down through the middle. Inside of it is the record of an odd natural history story.

There is a smooth channel down the center of it, a channel marking the path of a wood borer. This borer has planed off shavings as he went and has eaten them. They have made his breakfast, dinner, and supper. He has had no reason to worry about shortage of food.

It would look as though he were safe enough from other worries with an inch and a half of solid oak between him and any enemy.

One winter day a woodpecker came hopping up this tree trunk in his sprightly way. He was keenly hungry, for all the world about him was covered with snow, and most living things had hidden away beyond his reach. Life in the North was hard in such weather. Most of the birds had long ago gone south, where insects were plentiful.

This woodpecker tapped here, tapped there, and listened carefully. To be sure of hearing better he cocked his hammer head first on one side and then on the other. Perhaps a slight hollow sound showed him when he was over the wood borer's burrow, and a duller one that he had followed it to the place where the grub was resting at the time.

Then, of a sudden, he fell to work with his chisel-like bill. He worked hard, for oak is no easy wood to chip away. In the end, however, he reached his goal. Here in the side of the block you can see the well he dug. It strikes the tunnel of the borer just short of its end. It was here that the body of the grub rested. It was here the grub met death, for the woodpecker made one gulp of him that winter day when he was so hungry.

SECTION OF A TREE SHOWING A WOODPECKER DRILLING

One of the strange things about Nature is the way she arranges matters so they will balance, and keeps any one of her creatures from becoming too numerous. The borers, for instance, working inside trees, are beetles in the grub stage. They are grubs that will be beetles when they grow up. They are safe from being devoured by any bird or beast except this one, the woodpecker. But for him there might come to be too many borers. They might destroy the forests. It is the woodpecker's business to keep borers in check. He is the good fairy, the protector, of the trees.

Besides these borers that work deep inside the trees, there are others that live in the bark or just under it. One of these operates in that soft layer of material, called cambium, that is between the bark and the body of the tree. It is through this layer that the sap climbs up to nourish the tree. If the borer cuts this layer all the way around the tree its nourishment is shut off and it dies. Many of the big trees that you see standing dead among their fellows in the forests have been killed in this way. Sometimes whole forests are destroyed by these borers.

When Nature set out to develop a check on the borers she chose a bird, chiefly because the bird could get about the trees so well, but also because its bill could be developed for drilling. She found, however, that there were a great many changes she would have to make in this bird to cause it exactly to fit her purpose.

Its bill had to be hardened and strengthened. The bill of no other bird will cut its way through hard wood. Then it was necessary to get the woodpecker's head set on its neck more as a hammer is set on its handle. The heads of other small birds, if you will notice, are not set on in this way.

It was found that the feet of none of the birds suited them for the sort of walking the tree protectors had to do. These birds must be able to climb up and down the sides of trees, like telephone linemen. They must be able to grip into the bark and hold.

So Nature set about changing the foot of the woodpecker. The normal bird foot has three toes pointing forward and one backward. Feet to be used as grappling irons should be arranged differently. They should have more strength behind. So nature turned one of the foreward toes backward. They balanced against each other and bit into the bark like an iceman's tongs.

Even at that it was hard for the woodpecker to stand comfortably there on the side of a tree for half an hour while he drilled out a grub. It was a great strain on his muscles. After he had gripped the bark with his feet he needed a prop to hold part of his weight. He began using his tail for this purpose. The more he used it the stiffer the feathers became. The ends of them changed to sharp barbs that stuck into the trunk. The woodpecker came to have a tail that served its own special needs. The chimney swift also has a tail that it uses as a prop and grappling feet, but there are few birds with such tools.

But this was not all. The woodpecker needed one other instrument peculiar to its own sort of work—a harpoon with which it could spike and pull out these grubs hidden in the wood. Nature set about developing its tongue to serve this purpose.

The tongue of the woodpecker has grown to be a thing of surprising length. It is twice as long as the whole head of the bird. It is so long that the bird must have a place reaching all around the back of its head in which to wind up its tongue when it is not in use. The end of it is hard, stiff, and sharp. It has barbs on it like a fishhook.

Suppose the bird has drilled its way into the tunnel in which a borer is at work. It would need a quite wide space in which to open its beak and take hold of the grub. Maybe that gentleman has taken alarm and withdrawn a little way down the tunnel. What the bird does, instead of trying to grab the borer, is to spear it with his tongue. He can run that tongue out an inch beyond his bill, drive it into his victim, and drag it forth. It is a marvelous instrument for just this purpose; this tongue fits the woodpecker's special needs.

Thus equipped, this bird sets about its peculiar task. It is guardian of the tree trunks and keeps to its field, just as the oriole catches only the insects that infest the leaves and branches. In addition to the borers, it destroys the young of many moths which like to hide their cocoons in the crevices in the bark. The caterpillars come out of these cocoons and attack the young leaves in the spring. In the orchard the woodpeckers delight in catching codling moth grubs which are the arch enemies of the apple crop. They are everywhere useful.

There are two kinds of woodpeckers, the hairy woodpecker and the downy woodpecker, that stand at the top of the list in helpfulness. They are almost exactly alike in color, but the hairy woodpecker is about a third bigger and has different call notes and a different tattoo. The downy is the smallest of all the native woodpeckers. Both of these birds are very wide-spread. Both are light underneath, and dark on top, with mottles on their wings and a little splotch of red on each side of the nape in the male.

These birds stand at the top of the list in usefulness

because they work unceasingly, summer and winter, in their attacks on just the insects that do the most harm. It may be the height of the insect season when the trees are full of katydids and the meadows ateem with bees and butterflies, but these earnest birds stick to hard work and drill their dinners out of tree trunks just the same. That is their job and they will not be taken away from it.

The red-headed woodpecker, a very handsome and striking figure, is all fitted out for drilling, but it seems that he has been led to give up his proper task by finding an easier living elsewhere. It was probably because of his flashy clothes that he got into the habit of sitting on the top of telegraph poles. There were many grasshoppers playing around and he took to catching them instead of doing his proper work of digging grubs. Having yielded in this respect to the call of the easy life, it was but natural that he should be tempted elsewhere. Where formerly his purpose was to pick borers from the apple tree and worms from the apples, he began nibbling the fruit itself and at times has become almost as bad as the robin as a fruit eater. Unless he reforms he may some day be set down as an outlaw and any man with a gun may be invited to shoot him. This would quite spoil his showing off on a telegraph pole.

There is another of the woodpeckers that has been led away to a degree from the purpose for which he was created, although he still follows a quite useful life. This is the flicker, or golden-winged woodpecker, a very widely distributed bird and very handsome, mottled black and buff all over the back, with a gorgeous red band across the nape. a black crescent on its breast, and a glow of gold beneath its wings. The flicker, you will find, often flies up from the ground in front of you as you tramp about the fields. Why, do you suppose, is this wood-

THE FLICKER

pecker, this intended guardian of the tree trunks, down there on the ground?

To tell the truth, the flicker has developed the ant habit.

It acquired the habit, to be sure, up there in the trees. The trees are full of ants. It is there that the ants keep their cattle and they must go up for the milking. Aphids, or plant lice, are ant cattle and they live in the trees. The ants take care of their aphids, protect them against their enemies, and hide them away under ground that they may live through the winter. The aphids pay the ants in honey for the trouble they take. Many ants are not directly injurious to man, but aphids are among the worst enemies to plant life, and so the ants, in keeping them alive, are indirectly injurious.

The flicker undoubtedly got the habit of eating ants while they were climbing up and down the trees. Then, finally, he made the discovery that there were ants' nests down in the ground and that this sort of food could be captured much more rapidly by going to headquarters. So now the flicker goes down to the nest, pounds on the ground to arouse the ants, and when they come swarming out, laps them up. It has learned to use its tongue in a peculiar way to do this. It has acquired the knack of putting glue on its tongue. It runs it down the ant hole, gets half a dozen victims at a time stuck on it, and withdraws it. Repeating this act it soon has secured an ample, if highly spiced, dinner.

Another familiar woodpecker which has been led to abandon the admirable industry of its kind is the sapsucker, a very widely distributed species, splashy red about the head, yellow underneath, black and white on top, and with a broad white stripe up and down the wing, a mark which none of the other woodpeckers has.

It appears that the sapsucker originally must have

drilled in the wood for its food as do its cousins who have stuck to that hard working method of earning a living. In drilling its holes in the bark of the trees, however, the sapsucker made a discovery. It found that the sap which ran from certain of these holes was sweet, was good to drink. As time passed it also discovered that if it would drill out little basins beneath the wounds it made in the trees, these would fill up with sap and it might return from time to time and drink from them. Then also it found that ants and wasps belong to the bee family, which is very fond of sweets, and would come to these wounds for the sap. It might then easily catch them. Thus it worked out a scheme under which it could secure both food and drink merely by converting the holes that it dug into basins for catching sap. It was so easy to make these little basins in the soft outer bark that soon this woodpecker had yielded entirely to temptation and had become a sapsucker. It has now been getting its living in this way so long that its tongue with a spike on it for lancing grubs has disappeared. Instead it has a thick brushy tongue well fitted for lapping up sap.

This sapsucker also attacks the bark on the tender branches of trees, digging through it and eating the sweet cambium beneath. It cuts through the bark in such a way as to leave strips of it both up and down and across; in this way it often works out on the limb of the tree an embroidery-like pattern of very regular design. When one sees such a design on the limb of a tree in the orchard it is well to know that it was not made by a small boy's penknife, but by a woodpecker with a fondness for cambium.

The great spruce forests of the North which are among the most valuable timber areas of the world are subject to serious attacks by the bark beetle. It often happens that considerable areas are killed by it and much valuable timber goes to waste. When this happens, the greatest ally of the lumberman is a woodpecker which is different from the others in that it has but three toes instead of four. This three-toed woodpecker comes down from the Arctic, or from the cold regions in the high mountains, and sets upon these bark beetles and their grubs and devours great numbers of them. In this way a northern bird tribe is of great service to man creatures who live farther south, but need spruce for many purposes.

Woodpeckers are among the few birds that have an instinct for storing food. This storage is limited almost entirely to beechnuts and acorns, and California woodpeckers are the greatest of the hoarders. Their manner of storing acorns is unique. They drill holes, most frequently in the bark of living oak and pine trees, but often in dead trees, in fence and telegraph posts, in the sides of a barn, and in church steeples, and fit their acorns into these holes. They fit them so tightly that it is difficult for rats or squirrels to steal them. It often happens that the bodies of trees and posts are so closely studded with these hoarded acorns of the woodpecker that the holes are not more than an inch apart. When the time of food shortage arrives the birds return, drill into their acorns, and satisfy their hunger.

These California woodpeckers eat a larger percentage of vegetable food than any others, and it is largely the kernels of acorns. It is a fortunate hoarder, for it has no objections at all to worms getting into its acorns. When this happens it eats the worms, which it probably prefers to the acorns.

In prowling among the trees a close observer may sometimes see a very peculiar thing. He may find a great June bug with his back fitted snugly into a hole in the bark and his legs sticking out and waving quite helplessly since they can grasp nothing which is firmer than thin air. The man might well wonder how the June bug got himself in such a peculiar position, but if he watches long enough he will see a woodpecker come to get him and will then realize that fitting his victim tightly into this hole was the woodpecker's substitute for man's method of tethering a goat with a rope.

There are some forty-five species of woodpeckers in the United States and about three hundred and fifty kinds of them in the world. The biggest of all these in the United States is the ivory bill of the lower Mississippi and Florida, a gorgeous creature of the solitudes, now almost wiped out, whose wings have a spread of thirty inches.

All these birds have nesting habits that are peculiar to their kind. They use their ability to chisel into wood in making for themselves what are among the snuggest abodes in the bird world. The woodpecker's hole is likely to go straight into a dead limb or tree trunk for two or three inches and then to turn down the limb to a depth of from six inches to a foot and a half. This hole widens out toward the bottom and makes a cosy place in which the mother bird may deposit her eggs and bring up her family of from four to ten young ones.

The woodpecker builds himself a winter home in much the same way. He is by instinct such an industrious fellow that it may be merely because of his love of digging in wood that he builds a new home instead of continuing to use the old one.

So industrious is he, in fact, that he sometimes studs the side of a barn with many holes (thereby doing considerable injury to the farmer) for the mere purpose of exercising his bill. Another device of his of a similar nature is that of picking out a loose board on the house, or a loose sliver on an electric light pole, or even a bit of a tin roof, which by experiment he has found is a good sounding board, and of playing a noisy tattoo on it. He is by instinct a snare drummer with a good ear for time, and on such an instrument he beats a tattoo which seems to give him great satisfaction. Having located a good sounding board the same woodpecker returns to it day after day to practice his music.

Woodpeckers usually live summer and winter in the same region, but some of them shift back and forth for a few hundred miles with the seasons. The most to be prized of them all, the little downy, does not migrate at all, but may be seen through the winter mixing in a friendly way with the few other birds of the cold months. This downy woodpecker, the hairy woodpecker, and the three-toed woodpecker of the North, should be remembered as very valuable birds, never to be killed, but, on the contrary, to be especially favored and protected.

QUESTIONS

1. Describe the woodpecker's hunt for the borer on a cold winter day. The well he dug for his breakfast.
2. What do wood borers become when they are full grown? Describe the harm done by the borers that work just under the bark of a tree. What are their natural enemies? Show how the woodpecker protects the forest.
3. How has Nature strengthened the bill of the woodpecker? How is the woodpecker's head set on its neck? What change was made in the woodpecker's feet to fit it better for its task?
4. How did Nature give the woodpecker a prop that it might the better stand on the side of a tree? Finally, what sort of special tongue was given it? How does it work?
5. Each kind of bird has its special hunting ground. What part of the tree belongs to the woodpecker? What besides borers does it catch?
6. What are the two most useful woodpeckers? Describe them. Show how they stick to hard work.
7. Show how the red-headed woodpecker is drifting away from his job.
8. Describe the flicker. Why is it often seen on the ground? How did it get the habit of ant hunting? Why do ants climb trees? How does the flicker use its tongue in catching ants?
9. How did the sapsucker get its bad habit? What injury has it done to trees by drilling into the bark?
10. Where is the three-toed woodpecker to be found? How does he come to man's aid?
11. How do California woodpeckers store acorns?
12. Where do woodpeckers nest? Describe their winter homes. Their snare drumming.
13. There are many marvelous examples of changes that Nature has brought about in animals because of their special needs. Recall the wing of the penguin, the foot of the ostrich. Give any examples of such changes of which you happen to know.

CHAPTER XXI

THE DUCK

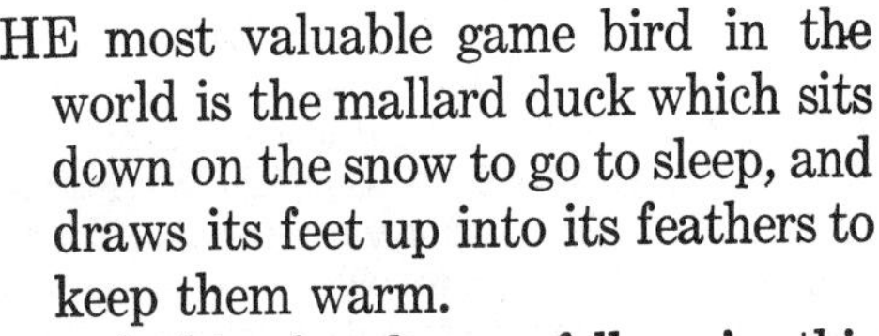

THE most valuable game bird in the world is the mallard duck which sits down on the snow to go to sleep, and draws its feet up into its feathers to keep them warm.

A big handsome fellow is this mallard, standing out as the most important member of the duck family. It is as numerous, in fact, as all the other ducks combined. It has furnished man, through the ages, with as much good flesh for food as all the other game birds put together. None, likewise, has provided more sport for the hunter.

In that primitive state of affairs which existed in America five hundred years ago, in most of Europe two thousand years ago, and throughout the whole world a little further back, there was a teeming abundance of waterfowl. Early man set his snares for these birds, and despoiled them of their eggs and their young. Civilized man shot them with his magic gun, served them on his individual table, shipped them in ton and carload lots to his market places, domesticated them and bent them to his will.

This king of the duck tribe, the mallard, breeds in the summer in a belt up North which reaches pretty well around the world. The mallard is at home at times in most places in the northern hemisphere and there are few people in the north temperate zone who do not know it.

In the autumn the mallard, together with most of the members of its tribe, moves south some two thousand miles, thus flicking its shadow over the houses of most of the human beings who live in North America, Europe, and Asia. Roughly, it breeds above the fiftieth parallel of latitude and winters along the thirtieth. In the spring it flies back again over that more than twenty degrees of latitude, visiting a bit at way stations en route.

The mallard duck is an international personage. In the western hemisphere it is usually a Canadian by birth, but a resident of the United States for more than two-thirds of its life. It crosses the border spring and fall with not so much as a how-do-you-do to immigration and customs agents. So important is it to both nations that it has brought about the making of a treaty which has been formally signed by those mighty nations, Great Britain and the United States.

The mallard duck has treaty rights. It may not be shot in either country except during certain brief seasons each year. Wild ducks may not be sold in the market places of either nation. Thus has the killing of ducks been limited to those shot by sportsmen at stated seasons. Thus has the sale of wild ducks been entirely prohibited.

In this way this game bird has been saved to future generations. A continent has taken action to protect one of its favorites, so that its presence through future years has been assured.

The treaty and the laws which came about because of the mallard duck also protect many other migratory birds. They stand out as the wisest bit of such action ever taken by great nations.

Not all American ducks are born outside of the United States. The lake region from Wisconsin to North Dakota, for instance, used to be the nesting place of great numbers of them, was spoken of as the paradise of ducks, and still has its retreats in which a great many of them nest. Settlers have come, however, have drained many of the lakes and marshes, and have built houses where ducks used to paddle. The same is true, but to a less extent, over the Canadian line in Manitoba. Much of the Canadian lake country, however, is still wild. Innumerable ducks still breed there. Still farther north along the west shore of Hudson Bay, sometimes reaching to the Arctic Ocean, are limitless solitudes suitable in summer as nesting places for ducks. This is the reserve duck hatchery of a continent.

The great mass of the ducks and geese that make their nests in this vast northern region find their chance of getting anything to eat gone when the water freezes in the autumn. They are driven south. The great drift is down through the Mississippi Valley to the Gulf Coast. Another much traveled route, however, is by way of the Great Lakes and thence to the Atlantic seaboard with Chesapeake Bay as a favorite resort. Much hunting has lessened the numbers of birds on Chesapeake Bay, but the time was when it was probably the greatest duck pond in all the world.

While masses move somewhat regularly north and south, individual groups of ducks and geese make strange journeys. There is, for instance, the white-winged scooter sometimes called the coot, which breeds in the Hudson Bay country, but splits its flocks, sending one-half of them to winter along the Atlantic Coast in the vicinity of New York, playing safe by keeping far out to sea, while the other half frolics three thousand miles away on the Pacific. The Ross snow goose, a very rare bird, breeds in the brief summer of the islands of the Arctic and, except for stragglers, visits only California for the winter. The blue goose, also rare, winters in one narrow strip near the mouth of the Mississippi and has been traced north as far as Hudson Bay, where, only recently, its nesting grounds have been discovered for the first time. Canvas-back ducks, favorites with those who take much delight in their food, breed along the lakes of Saskatchewan, and winter along the coast between Connecticut and South Carolina. The Canadian goose, noble cousin to the duck and among the biggest of water birds, breeds along the Canadian border with the mallard, and when it used to be more plentiful, spread out in season over most of the United States.

When settlers first came to the United States they found what seemed to be inexhaustible supplies of ducks and geese along the Atlantic Coast. It was three hundred years before they seriously reduced their numbers. When they traveled west, however, they were astonished by the masses of waterfowl. The lake and prairie country along the Canadian border gave the birds just what they wanted as a breeding place, and they flew back and forth to the Gulf of Mexico. All the waters between knew them in incomparable abundance. Through the centuries they had done much to make this region a happy hunting ground for the Indians, who were too few to reduce their numbers. They helped much in keeping early settlements supplied with food. Only dense populations of modern men with high powered guns have been able to overcome their huge numbers. Modern civilized man, as is his way, had quite upset the balance of Nature and was threatening the destruction of the waterfowl until governments stepped in to save them.

Not that they will ever thrive as of old, because farms now cover their former breeding places and rice plantations are crowding them out along the Gulf Coast. Adjustments of the shooting seasons, however, may be so made as to keep them quite plentiful.

These waterfowl in the winter time are thicker along the coast of Louisiana, between the mouth of the Mississippi River and the Texas border, than in any other region. Here there are great reaches of marsh lands with frequent lakes and an abundance of food to their liking. Farther west, along the Texas coast in the vicinity of Galveston and Corpus Christi, there are many waterfowl, and some of them reach down the coast as far as Tampico, Mexico.

It is this narrow strip that, in the winter, holds most of the waterfowl. It is here that sportsmen gather for shooting, during the season in the autumn when it is still allowed. It was to this region of old that the market hunters used to go to slaughter ducks and geese without numbers to be sent to the cities of the North for sale. It was stopping this market hunting and stopping all duck shooting in the spring that were the chief helps in giving these birds a new chance to multiply.

In the movement for the protection of waterfowl in the United States a number of surprising developments have occurred. In Oakland, California, for instance, a lake of salt water, covering nearly a square mile, was set aside as a water bird refuge. Water birds might come there and be safe. It is a strange thing that bird life soon comes to know about these safety zones and flocks to them. This lake in the center of a busy city soon came to be thronged with waterfowl. Often two or three thousand of them may be seen swimming on its surface or sunning themselves on the lawns that surround it. Automobiles hurrying by, foot passengers going along the walks in no way disturb these wild mallards, pintails, teals, and coots. They realize their safety and make the most of it.

Palm Beach, Florida, converted itself and the land for a mile around it into a bird sanctuary, and in this zone of safety waterfowls are as tame as chickens in the barnyard, but just outside of it they are as wild as partridges in the field.

The government has established scores of sanctuaries in which birds may not be disturbed—Pelican Islands, Florida; Bretan Island, Louisiana; lake regions in North Dakota; breeding grounds at the mouth of the Yukon, in Alaska; a home for gulls in the Hawaiian Islands, and many others. Private individuals have bought tracts and made them safe for the birds. One such tract, for instance, contains fifty thousand acres and lies at the mouth of the Mississippi River in the heart of the duck winter home. The waterfowl are coming at last to get a bit of kind treatment to make up for the outrages that have been committed against them in the past.

The scientists put the ducks, geese, and swans together in this waterfowl family. Anyone can see that they are very much alike. Their chief difference is in the length of their necks. The swans have necks longer than their bodies, some of them with as many as twenty-five joints in them. The geese have fewer joints in their necks than the swans, but more than the ducks.

The chief division in the duck groups is into river ducks and sea ducks. It does not take much knowledge to tell to which of these divisions a given specimen belongs.

RIVER DUCKS TIPPING (BLUE WING TEAL)

If the specimen is dead one has but to look at its back toe. All of these birds have three webbed toes in front and one lesser toe sticking out behind. If that toe is webbed the specimen is a sea duck. If it is not webbed it is a river duck.

If the specimen is alive and in the water it is equally easy to tell to which group it belongs. The river ducks eat off the surface, or in shallow water by "tipping." They put their heads under the water and tip their tails up, but they do not dive. The sea ducks, on the contrary, go clear under, preferring to eat off the bottom where the water is four or five feet deep. Some of them dive to surprising depths; "Old Squaws," for instance, have been caught in gill nets in Lake Erie one hundred and fifty feet down, and there is on record a case of a king eider feeding off the Pribilof Islands that was known to have dived to a depth of two hundred feet.

The test of two captive ducks in a pool to see to which group they belong is very interesting. A mallard and a canvasback, cooped up in this way, act very differently when fed. If grain is thrown on the water the mallard begins gathering it in and eating it as soon as it has struck the surface. It is a river duck. The canvasback, however, waits until it has gone to the bottom, and then goes under after it. By dint of paddling with its feet it holds itself under for some time while it picks up the grain. It is a sea duck. Thus it appears that, among the waterfowl, there is a division of the food field just as there is among the birds of the trees.

The river duck is more a land animal and more a vegetable eater than the sea duck. It even goes ashore and eats grass. The goose is more a grass eater than the duck, often grazing on land with cattle. In the water the goose and the swan have feeding habits much like the mallards and other river ducks. They feed in shallow waters by tipping.

The mallard is not only the most valuable, but the most numerous of the ducks. It is a lusty fellow and raises a big family. Mrs. Mallard lays nine or ten eggs, sometimes fourteen. She takes good care of them and the danger to them is not great up there in the North. Many members of the family may be lost in the long journey back and forth to the South without the numbers, through the years, becoming less.

One mallard, the greenhead of the Mississippi Valley and Gulf Coast, is less common in the East. In the North it does not breed east of Hudson Bay. It has a cousin, however, that does. This is the black mallard. Its range is up and down the north Atlantic Coast. It is like the western mallard except in color and disposition. It is more difficult to tame than the greenhead. Take the eggs of a wild greenhead and hatch them under a tame duck and they become a part of the flock. Do the same thing with black mallards and the first time the little ones get out of the coop they take to the brush. With careful guarding for several generations, however, even the black mallards become tame. There are now great flocks of domesticated black mallards in certain eastern states.

Almost all the tame ducks the world around come from the greenheaded mallard. The tame geese come from a wild goose much like the Canadian goose. It was not the Canadian goose that was first tamed, of course, for geese had lived with man—European and Asiatic—for several thousands of years. It was the cackling of geese, as you will remember, which once saved ancient Rome from being surprised by the barbarians from the North, and that was something like two thousand years ago. Geese have long been and still are an important thing in the peasant life in Europe.

One member of the duck family that stands out because of its odd habits is the wood duck. It has formed the habit of living in trees and its feet have become fitted for that mode of life. It is an unusually long-legged duck and lives in warmer countries than most of its kind. The wood duck is held to be the most beautiful of American ducks, combining white, brown, green, and chestnut, and boasting a strikingly speckled breast.

This wood duck comes up from the South. one of its favorite nesting places being the tree-fringed borders of the Potomac River near the national capital. The wood duck searches carefully for some cavity in a tree and there makes its nest. It is a strange thing that the number of eggs that it lays depends upon the size of the cavity it has found for a nest. It may lay no more than five or six eggs or it may lay as many as eighteen.

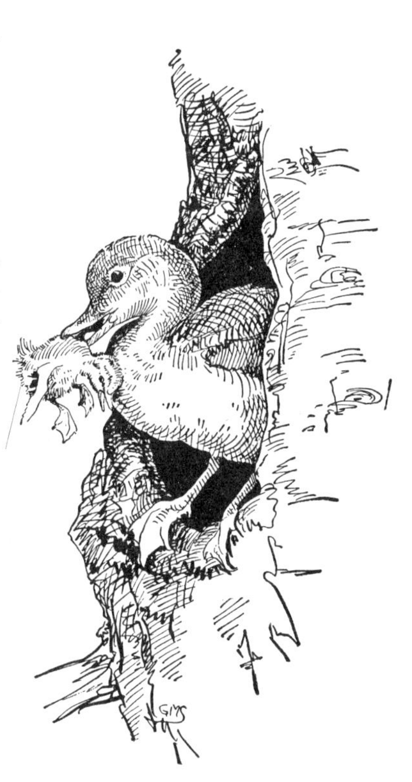

WOOD DUCK AND YOUNG AT NEST

If it happens that one of these wood ducks finds a good hollow tree for making a big nest a mile away from the water it lets itself in for a great deal of work. It produces this large family and, having done so, it is forced to transport it to the stream. This the parents do by taking the young birds one after another in their bills, very much as the old cat carries her kittens, and flying away to the stream with them. Arriving at the stream each little duck is dropped into the water and scampers quickly into the weeds to hide. The father and mother fly back and forth to the nest until the whole brood has been brought to the water. They then call the hidden birds, and, though they have never been in the water before, they immediately go darting about catching food.

Another very peculiar duck, and at the same time a very valuable one, is the eider duck of the North. The eider duck has been for many centuries a good friend of the people of northern Europe. In the western hemisphere it is a favorite of the people who live in Iceland and Greenland. Northern Europeans prize this duck very highly and take very good care of it. Icelanders are more careless of their eider ducks, but still they can be seen to advantage among that people, and the peculiar methods in which they are made useful may be observed.

The eider duck is protected in Iceland and is therefore not afraid of human beings. It comes, in fact, and makes its nest on the roofs of the cottages. When the eider duck has completed its nest there is one peculiar idea in its mind to which it sticks doggedly. It must have eleven eggs in its nest. The Icelanders know that the eider duck is determined on eleven eggs, and so, beginning in the early spring, they watch the nests and see to it that none of them ever comes to have that many eggs. They take the eggs out and eat them. Thus the eider duck is forced to keep on laying eggs. It does this for some two months, much to the profit of the Icelanders.

Then, at about the first of July, the Icelander realizes that the duck must be given a chance. It is allowed to round out its eleven eggs. Then it quickly stops laying and goes to sitting.

These eggs, however, are not the only profit that comes to the people of the North from the eider duck. That choicest of bed covering, the eider-down quilt, results from the fact that these northern people take their toll of down from the nests of these ducks. When the eider duck (or any of the other ducks for that matter) is completing its nest it is very particular about its lining. The nest, close to the eggs, must consist of a thick blanket of down. In making this blanket the duck plucks the under feathers from its own breast. These are woven together, perhaps with the added mixture of a little straw, until a very comfortable blanket is made. It extends far up on the sides of the nest and, when the mother duck finds it necessary to leave to get something to eat, she pulls the edges of her blanket over the eggs, thus protecting them from the cold and, likewise, from the prying eyes of any sea gull which might have the idea of gobbling them up.

The Icelanders figure, however, that Mrs. Eider Duck uses more down than she needs in lining her nest and that they may take a bit of it from these nests. These downy feathers which the northern bird plucks from her living breast are said to have a different quality from the feathers plucked from the dead eider ducks. In the trade they are called "live feathers" and bring higher prices than the others. The Icelanders take an average of one-sixth of a pound of down from the nest of each duck, and harvesting this is one of the industries of the island.

QUESTIONS

1. What is the position of the mallard duck in the duck family? What is its rank as a game bird? What was its position in the distant past?
2. Where does the mallard breed? Where does it spend its winters? How far does it travel in its migrations?
3 Why was it necessary to make a treaty about ducks? What are their treaty rights? What have been the results of this treaty?
4. Describe the duck breeding grounds in the North. What forces them to the South? Show on the map some of the odd journeys that certain ducks take. Trace the journeys of the coot, the snow goose, the blue goose, the canvas back.
5. Tell about the abundance of waterfowl when white men first came to America. What help were they to settlers?
6. Where are waterfowl most abundant in the winter time? Describe the shooting. The sale of the kill. What effect is control of shooting having?
7. What is a bird sanctuary? Birds soon learn where they are safe. Describe the ducks of Oakland. Those of Palm Beach. Where are there other havens? How do they act outside the sanctuary?
8. What are the three cousins of the waterfowl family? Which have the longest necks?
9. How can you tell river ducks from sea ducks? When dead? When alive? How deep have ducks been known to dive?
10. Which kind of ducks do geese and swans most resemble? How do they feed?
11. How large a family does Mrs. Mallard raise? Where is the black mallard found? From what tribes do tame ducks come? Has the goose lived long with man?
12. Describe the peculiarities of the wood duck. Where does it nest? How does it get the little ones to the water?
13. Where does the eider duck live? Show how it is led to produce eggs. How is eider-down secured from them?

CHAPTER XXII

THE WOOD WARBLER

SOME day you may witness this sensation of the woods. Suppose that you are strolling through the forest solitude on a May or a September afternoon. There is little movement other than the frisking of an occasional squirrel and little noise beyond the cawing of a chance crow. All is peace, quiet, silence.

Then, of a sudden, bedlam breaks loose. From silence and solitude the forest in a twinkling becomes thronged with innumerable living creatures, each full of its individual sort of sweet chatter, each contributing to the sound that fills the woods with a pleasant hum of life, each active beyond belief, darting ceaselessly about like a rocket in the dark woods, and so adding its bit to an interweaving of motion and color that is without limit.

Of a sudden it has happened that a rabble of birds has burst into this forest. The strange thing about it is the fact that this is not a flock of birds all of the same kind, like the blackbirds that sometimes descend upon the barnyard. There are woodpeckers among them, creepers, chickadees, nuthatches, and, above all, warblers. Most of them are warblers. Yet not even the warblers are all of the same kind, but of many different varieties, wearing different costumes. There are yellow warblers, blackpolls, Myrtles, redstarts. There may be warblers of twenty different species, each with a different dress and different habits. This does not seem like a group of individuals that belong together out for a party, but a mob made up from everywhere and therefore on some strange mission.

One thing you will notice about them, however, is the fact that they are all woods birds. They all belong to those shy groups of tiny birds that hide away in the forests and are not very plentiful around barn-yards and in open fields. They are of those forest and thicket tribes that are very numerous, but little known. They are of those hordes that are beyond counting that mysteriously come and go. They are colorful, suggesting the tropics.

Hosts of warblers live in Canada in the summer-time and in South America in the winter-time. This is probably such a group. If it is May that you see them, they are going north. If it is September, they are southbound.

Tiny creatures that they are they are great travelers, and powerful on the wing. It is not often that they migrate thus through the woods. They are more likely to be seen in the late afternoon in flocks that have no end, flying high overhead. This group, however, is breaking its journey for a picnic. It has come down to travel for a while through the trees, to frolic and feed as it goes. Though its members seem to be darting in every conceivable direction you will see, if you watch closely, that the flock makes progress one way. If you will wait long enough it will pass and the early quietness will again settle down on the woods. This wave of birds will have swept through and beyond your part of the woods.

The fact that so many different kinds of birds joined

together in this group shows another thing about bird nature. Birds are sociable beings. They like their fellows. Whether they are of the same species or not they tend, except at the nesting season, to flock together. When nesting time comes most birds pair off and each pair tries to lose itself from all the world. But when the young are grown, they are likely to throw in their lot with their fellows and live, for the rest of the year, in flocks. This mixture of different flocks, however, is likely to be more marked in the woods birds than most others.

This flitting through the woods is, however, but a suggestion of the movement of birds at migration time. To him who would like to get an actual look at it the possibility may exist. To anyone who can make an opportunity to spend a night, late in September, on some high point—a lighthouse, a wireless tower, the Washington Monument, a ranger's lookout station, even in a church steeple or other tall building—a great, unknown world may be revealed. He may get high enough to look in on what is happening up there in the blue where there is supposed to be a great stillness and absence of life. He is pretty sure to find, at this season of the year, that he has tapped a sea of activity of which the quiet world below never dreams.

On September nights the shy and timid birds are going south. The bold and vigorous fliers such as the doves, and those that feed on the wing, as the swallows, fly by day, but the timid woods birds and those that feed at or near the ground are likely to travel by night that they may be more safe and may use the daytime for foraging.

The air up above is full of these night fliers.

On one September night some one with a telescope set it upon the moon and counted the birds that passed between him and that bright globe. Through the night they came along, passing through that small circle of vision at the rate of a hundred an hour. If a hundred an hour went through this small slit how many, would you suppose, passed through the arc of the sky within the range of the eye of this one person? Compute that number and multiply it by the number of such arcs between New Jersey and Oregon and you will have some idea of the vast number of birds flying south this night up there where it is supposed to be so quiet.

Sometimes tragic proof is to be found in plenty of the movement of birds. During September there are many mornings when a bushel basketful of warblers can be picked up at the base of the Statue of Liberty in New York Harbor, or at the base of the Washington Monument, there on the Potomac, or at Sombrero Key Lighthouse, off the tip of Florida, or at any of a score of other places where high structures penetrate the bird streams that are flowing and where, perhaps, there are lights to blind the fliers. These frail little warblers are specially apt to meet death by flying against such objects.

It is odd that such frail little birds should undertake these long and perilous journeys. They are in reality birds of the tropics and spend three-fourths of the year in dense forests, in tropical countries south of the United States ranging all the way to Brazil. Their gorgeous plumage, in the first place, proclaims many of them birds of the tropics. Their feather coats are light and thin and of little value in keeping them warm if it happens to turn cold. They know this and wait until very late spring to come north. Yet even then there have been occasions when, late in May, rain and sleet storms in such northern states as Minnesota have overtaken the warbler flocks and killed them by the millions.

Countless numbers of these tiny tropic birds feel the urge each spring to go north. Perhaps it is because of their desire to be alone during the nesting time coupled with their knowledge of the presence of great open spaces in the North. Perhaps it is because of something in their past, the fact that the nests of their ancestors were in the North, and that they themselves were born there, that they have the instinct to return to their birthplaces to raise their own young.

Anyway, all the warblers of these breeds of all the tropics come north in the late spring and go back south in the early autumn. The tropics lend these countless flocks to the temperate zone for a brief three months. Then they are returned and with interest. They not only spread through the solitary forests that fringe the Yukon, and those of Newfoundland, far away to the east, but certain of them find summer homes in the United States, all the way down to the Gulf. There are summer warblers in all the forests of America.

The warblers come at just the time when they are most needed. Summer-time is the insect time and, in the adjustment of Nature, the warblers have their special piece of work to perform. They are insect catchers and their special field is the forest. The woodpeckers are busy on the trunks of the trees, boring for beetles and puncturing moth cocoons, but there are innumerable other plagues of the trees. The worst of these are the tiny aphids or plant lice, too small for the attention of most birds, and the eggs and larvæ of bigger insects.

Each different kind of warbler, since there is such a careful allotment of work in the bird world, has its special

OVENBIRD WALKING

part of the tree to look after. One kind picks constantly about the trunk and big branches of the tree. Another variety is working just as hard among the twigs and leaves. Yet another hustles its food on the ground beneath the trees. Altogether they render a stupendous service to the forests and it is hard to say how the trees get along without this loan from the South.

Another strange thing about the warblers is the fact that very few of them warble at all and these very rarely. The name of warbler is undoubtedly a mistake. It is likely that it came about in this way. In Europe there was a tiny woods bird known as a warbler. It did actually sing and deserved its name. When Europeans came to America and saw similar little birds flitting about among the trees they called them warblers. They are not, however, even of the same family as the little brown birds that are known as warblers in Europe. Neither are they related to the mocking bird and the thrush, master singers of the thickets. They are shy little things that do sometimes sing at mating time, but few bird lovers ever hear them do it.

There is the case of the ovenbird, for instance, which is "the warbler that walks on the leaves." It walks man-like, one foot at a time, in a mincing sort of way, scorning to hop as do many birds. Most of the warblers are tree-top birds which is one reason why, though there are such numbers of them, they are so little known. The ovenbird is the member of the family that takes care of the insects that are on the forest floor. It got its name of ovenbird from the fact that it builds a nest of grass with a cover over it that looks for all the world like an old-fashioned oven.

The ovenbird is better known than most of its cousins because of its well-known call. Every country boy has heard this bird calling insistently, "teacher, *teacher,* TEACHER, TEACHER, **TEACHER.**" It is a dainty little bird with a bright, olive-green back, ermine breast, and dull reddish-gold crown, sometimes likened to a tiny partridge because it lives on the ground.

For a long time nobody knew that the ovenbird was a most rapturous singer, that it had any other than this "teacher" call. For that matter, it was a long time before anybody knew that the noisy and quarrelsome jay, at the mating time, sings softly and sweetly to its mate. The hermit thrush of the pine forests, a contender for honors as the sweetest singer in all the world, was long thought to be a silent bird. Careful observers finally found these birds at mating time, when their souls were poured out in music, and discovered that the plain, shy creatures possessed unsurpassed, hidden talents. Greatest of all transformations, perhaps, is that which takes place at mating time and changes the tuneless reed bird into the melodious bobolink of the meadows. Finally, it was discovered that the shy little ovenbird also well deserves the name of warbler which had been given it by mistake.

The ovenbird sings in the evening twilight only during the mating season. It leaves the bed of the forest, for this is a flight song—a wild, rapturous melodious outpouring, a thing of glory done in secret.

A member of this group, a shy little woods bird, is the black and white warbler which, it seems, never warbles at all. Shyly it flits about the trunks of trees, reminding one of the creepers which, however, are of a different family. It seems to appreciate the advantage of a black and white contrast, as it is strikingly striped from head to tail in these colors. Daintily this little bird each summer picks insect larvæ from the trunks of trees from the

Gulf of Mexico to Hudson Bay, an act which is much to the advantage of the trees.

Quite different again are the habits of the redstart which sits quietly on a twig in its tree and waits for an insect

THE MARYLAND YELLOW THROAT

to pass near, darts out, catches it in the air as do few warblers, returns to its perch, and devours it at leisure. The male redstart brings with it the gorgeous plumage of the tropics, a combination of black and yellow. These radiant colors, however, are worn only by the males, and then not by the youngsters, for it may be set down that a redstart decked in these colors is at least three years old.

The Maryland yellow throat, with its black face and olive-green back, is a familiar warbler of the thickets. It flits about shyly and twitteringly, but in the eventide sings a flight song. It is the little Maryland yellow throat that is most often badly treated by that low and lazy scoundrel of the feathered world, the cowbird, which lays its eggs in this timid creature's nest. The young cowbirds that are hatched out develop into greedy bullies that get all the food the parent birds bring and so starve the little yellow throats to death.

The biggest of the warblers is the yellow-breasted chat, sometimes called the yellow mocking bird, a shy creature of the underbrush, of southern and central United States. Half its under body is yellow, and half white, while above it is olive-green. It builds its nest near the ground and in the quiet stretches chatters away in a manner which suggests the catbird since, one after the other, it seems to imitate the caw of the crow, the bark of the fox, the quack of the duck, the whimper of a puppy, and the rattle of the kingfisher. The chat also has a wing song of no mean quality.

Altogether there are about two hundred different kinds of these wood warblers, thirty or forty of which visit the United States, all exquisite, dainty birds, whose acquaintance is well worth making. They are strictly American birds. The students of birds, the ornithologists, point out the fact that they have but nine primaries, big feathers, on their wings, while the warblers of Europe have ten primaries. It is upon such differences as these that birds are divided into families. Thus are the warblers of the two hemispheres shown not to be related.

These warblers are everywhere in America in the summer-time. All the woods are full of them, yet, strangely, few people know them. Their shyness, and their homing in the forests, keep them hidden despite the fact that they often wear brilliant colors. Whoever takes the trouble really to make the acquaintance of a warbler, however, is likely to find himself well paid, as is one who studies the masterpiece of some old painter of miniatures or commits to memory some little jewel from a great poet.

QUESTIONS

1. Describe the appearance of warbler travelers in the woods. What kind of birds are in this throng? What is the explanation of their appearance?
2. At what seasons are birds sociable? When do they want to be alone?
3. Where should one go, and at what time, to see birds during migration? Have you ever watched them on the move? What birds travel by day? Why do most of them travel by night?
4. How many birds may pass across the face of the moon in an hour? In a night? How many, do you imagine, pass within view of your own housetop? Can you conceive the number that cross the fortieth parallel in a night?
5. Explain the death of many warblers due to striking high objects.
6. Where is the real home of the warbler? What facts show it to be a tropical bird?
7. Why do warblers come north? How far north do they go? How long do they stay?
8. What purpose do warblers serve while in the North? Explain their division of work in the forest.
9. Do warblers warble? How did they come by their name? Are they related to those master singers, the thrushes?

10. Most of the warblers are tree-top birds. Describe the ovenbird, which is a warbler. What part of Nature's house does it care for? Tell of the occasion when it soars and sings.
11. Describe the black and white warbler. On what part of the tree does it work? What is the redstart's method of hunting? Which redstart wears the gaudy clothes?
12. What other bird is the Maryland yellow throat like? How does the cowbird impose on the yellow throat?
13. Describe the chat and its imitations.
14. How many kinds of warblers are there? Do you ever see them? Why are they so hard to find?

CHAPTER XXIII

THE SWALLOW

THERE is an Indian legend which holds that the swallows are children transformed into birds. Ages ago, it says, groups of children played at making dirt houses at the foot of a cliff. They piled damp earth on their feet, tamped it down, then pulled the feet out, leaving caverns beneath.

Always they looked up at the clean front of the rock and thought what a wondrous game it would be to build mud houses in the crevices, and to plaster them against this high wall where no lumbering bear could come to knock them down. They wanted so badly to do this thing that, in the end, they were changed into swallows and have ever since been playing the game of mud house, out of reach of all harm.

That the legend is true, that these are the children birds, the Indians hold, is proved by the fact that all the swallows dart, circle, frolic, and play all day long, up there in the blue, as grown ups would never think of doing.

The swallows, in fact, are as much at home in the air as any birds that fly. They spend almost all their waking day on the wing. They circle and glide in the air with the ease and freedom of a fish in the water. They are considered among the most graceful of birds. They lazy along, seemingly without effort or purpose, go zooming up, volplane toward the earth, turn sharply to right or left. There is an unbelievable speed in their blade-like wings.

These birds spend so much time on the wing, in fact, that they find almost no use for feet. They rarely touch their feet to the ground except on one occasion. That is the time when they go down to a wet place to get balls of mud with which to build their houses. They never use their feet for hopping about in trees or on the ground. They use them only in perching. Whenever they move about it is their wings that carry them. Because of this manner of life their wings are highly developed and their feet are small and weak.

Of all the feet of birds, theirs are among the weakest. Of all the bills of birds, also, theirs are least fitted for heavy work like that of cracking a grain of corn, or crushing the life out of a sturdy insect.

They do not need strong bills. That marvelous instrument of the bird with which it defends itself, feeds itself, builds its house, and makes its toilet is, in each case, fitted to its peculiar uses. The crossbill uses its beak as a gouge with which to pry seeds out of the pine cone. The crow may crack an acorn with its beak. That of the snake bird is a lance; that of the duck serves as a sieve; the snipe has a probe; the humming bird a syrup straw; the woodpecker a chisel; the oriole an awl. The swallow puts its bill to none of these uses. It may shape the soft mud of its nest with it. Other than that it is used only for purposes of opening and closing that trap which is its mouth and in which it catches its insect food.

The chief feature of the mouth of the swallow is the fact that it opens from ear to ear. It is a very large mouth.

Further, it is a sticky mouth. These two facts help the bird in making its living.

As a matter of fact, the swallow is by no means as playful, darting about the sky, as it would seem to be. There is an element of business in its journeyings. It gets its living while it plays. It is making use of its master wings and its wide-hinged mouth for purposes of catching insects. The air up there where it works is its special hunting ground. We have seen that the birds have carefully divided the insect field and that each has its place to hunt. The woodpeckers dig grubs from the bark of trees; the creepers pick them from the trunk; the orioles work among the branches; the ovenbird searches on the ground beneath the trees; the redstart waits on a twig until some flying insect comes past, launches forth, and captures it.

The swallows do not poach on the preserves of any of these. They unhinge their wide, sticky mouths and scoop up the small insects. All the insects that touch their mouths are there to stay, for the glue holds them. They catch the insects where they have left their food plants, or hiding places, and are cruising about in the air. The fly family is the largest item of swallow food, but bugs, ants, wasps, beetles, and numerous other insects are eaten. The swallows do much good in devouring annoying and destructive insects.

It is a sort of pot shooting that the swallows do among the small insects, but they pursue individuals, too. There are many bigger flies in the air. These flies hover over the surface of quiet waters and swallows often may be seen skimming such waters. If there is a water hole in the pasture where the cows come to drink, where they like to stand knee deep, many flies are likely to gather to plague them. The swallows know that flies follow the cattle and they, also, trail along. For the same purpose they frequent barn lots where horses are kept. The relief that they bring to tortured livestock is such that horses and cows, if they went in for such things, should erect a huge monument, surmounted by a swallow, rampant.

Except in the case of the tree swallow which fattens on berries during the southward migration, swallows almost never touch any food other than insects caught on the wing. They are very wise birds in the business of making a living, since they employ their excellent wings to keep them always in a region where there are insects. They never appear in the North until the occupants of the insect eggs of the year before have hatched out in spring and taken to the air. In the autumn, when insects begin to get scarce, they start south. They loaf along to the Gulf Coast, to Mexico, to Central America, to Venezuela, to Brazil. In the tropics, where there is perpetual summer and abundant supplies of insects, most of them spend what is the cold season in the United States. Then, in the spring, they come north as the insects take to the air, traveling by day, feeding as they go, some of them never stopping until they have reached Greenland and Alaska.

Swallows are a very distinct family and have no near relatives. It used to be thought that the chimney swifts belonged to the same family, but when the two were compared, they were found to be far apart. They have, for instance, a different number of primaries, big feathers, in their wings and tails.

There are half a dozen different swallows. There is

the tree swallow, for instance, with lustrous green back and white breast, that goes no further south than Florida to winter; this more northern bird is the first to appear in the spring. These birds are very fond of stringing themselves like so many beads along a telephone wire. When not nesting they spend the nights in great flocks in some marsh and, at daybreak, scatter out over the country for scores of miles, gathering insects all day long. In the dusk they strike back to the marsh. Millions may be seen coming in after sunset. A good idea may be gathered of the incomparable speed of the swallow by watching some individual that has stayed overlong before starting to the roost and therefore turns on full steam during the trip home.

Many of these tree swallows no longer deserve the name. Their association with man has brought about a difference in their manner of life. They used to nest in hollow trees, often in the deserted holes of woodpeckers. Then man came along and began putting up bird boxes. The tree swallows like these so well that they have largely given up their former manner of life and now live almost exclusively in homes provided by man.

The cliff swallow, however, has come even more strongly under the influence of man. When the white man first came to America he found in many parts of the country that wherever a sheer cliff presented its face it was likely to be quite covered with the nests of these swallows. Often its front would be covered with their mud-built nests as thick as they could be placed. It would appear to be a many-storied bird tenement.

Then man began to scatter his buildings all about.

The wall of a barn had all the advantages of a cliff, and possibly more. The thing that these builders in mud wanted more than anything else was protection from the rains that might melt their homes. For this the side of a barn was likely to be better than a cliff because it had overhanging eaves which kept the rain off. So the cliff swallows got the habit of building their mud nests close up under the eaves of barns and houses. They have almost ceased to be cliff swallows, except in uninhabited parts of the country, and have become eave swallows. They are coming to be generally called eave swallows.

A careful observer of a group of these swallow nests under the eaves saw a thing done that shows a surprising intelligence among these dwellers in close quarters. One day he noticed that the birds were bringing mud and sealing up the narrow opening of one of these nests. He wondered why this was being done, but, not wanting to disturb the colony, waited until the season was over to investigate. He then broke open the sealed nest and found in it the remains of a dead swallow. This mother had died on her nest and this was their effort at sanitation.

It is odd that there are so many groups of swallows, very much alike, all living together in the same barn-yard, yet each keeping to itself. The barn swallow, for example, never mixes with the eave swallow. His nest is built on a beam inside the barn. It is a bit different in architecture from that of his cousin under the eaves. Neither ever visits the other. Neither makes any concession to the other in his manner of life.

The barn swallow is likely to have a warmer place in the heart of the country boy than any of the others. This is the swallow that he comes to know most intimately, largely through playing in the hayloft on rainy days. The barn swallow twitters sociably, comes and goes without concern, pays little attention to his human associates. All the swallows are fond of conversation. They twitter constantly and very pleasantly.

BARN SWALLOWS

He is the handsomest of them all, most gaily robed, this barn swallow, with a splash of reddish buff on his breast, and a tint of it throughout his coloring. His forked tail is another feature that sets him apart, for none of the others have tails so deeply divided. Its excellence as a rudder has a good deal to do with his very skilful flying.

Then there is another swallow civilization at home in bluffs along streams and roadways. The bank swallows, smaller and duller colored birds, dig holes in these banks and build their nests in them. Like all the other swallows they, too, are sociable and live in colonies. One bank may provide homes for hundreds of them and the comings and goings of them suggest those of a hive of bees.

Finally, there is the purple martin, which is also a swallow, the biggest of them all, which travels from ocean to ocean and from Brazil to Alaska, and is one of the best known. Purple martins used to make their homes in hollows in trees or crevices in the rocks, but that was before white men came to America and began putting up houses for them. Even before that time the Indians had found out the martins' liking for ready-made houses, and often hung out gourds in which they nested.

The Indians had a reason for this. The martins are quite warlike birds. They give battle to any hawk, buzzard, or crow that comes near them. The Indians used to kill deer and hang the skins and the meat out to dry. Buzzards, as is their way, came to feed on this flesh. The martins, however, would drive them away. They were the buzzard watchdogs for the Indians.

The martins serve a similar purpose in nesting near a farmhouse. They give battle to the hawks and thus make for safety among the poultry. The negroes of the South still use the same style of gourds that the Indians once used, swinging them to a pole, and will tell you that the martins keep the hawks away.

Since Colonial days martin houses on the tops of poles have been common in the United States. These houses may have many compartments, and every one, because of the sociability of the birds, is likely to be occupied. There used to be an inn beside one of the ancient post roads of New Jersey and in front of it stood a martin house of five stories. The proprietor for thirty years wrote on its supporting post a record of the dates of arrival of the martins in the spring. A perusal of this record was an interesting experience to those who stopped at the inn. One day the proprietor hired a man to paint a fence and, in the excess of zeal, while nobody was looking, he painted the support to the martin house as well, and obliterated this most interesting migration record.

Since the English sparrow was brought to the United States the purple martin has had to do a great deal of fighting to keep its houses. The martins and the sparrows are at it all the time. The martins, it seems, are a little more than a match for the sparrows, but, nevertheless, are much worried by them, and it is likely that some martins fail to find nesting places because of them.

Not many people in this generation have ever seen a purple martin nest anywhere but in the homes that man builds for them. In fact, the birds have come to depend so much on man for nesting places that, if they fail to find them ready prepared, they will go on a strike and not nest at all. They are among the swallows that know man best. All of this group, in fact, except the bank swallows have given over more or less completely their natural nesting habits and have come to live in places provided by man. The swallow has flourished under the influence of man and has come to be as intimately associated with him, probably, as any untamed member of the bird world.

CLIFF SWALLOW AT NEST

The nesting habits of all these swallows are very much alike. They usually lay about four or five eggs, which shows that they have about an average chance to survive. They feed their young entirely on insects. They are so at their ease in the air that one may sometimes see a parent and a young bird, after it is able to fly, meet aloft, and hover there with bills together, while food passes from one to the other.

All of the swallows are graceful, beautiful, happy birds that have much endeared themselves to man. All are, in fact, very useful to man. They consume enormous quantities of the insects that most plague him. Many orchardmen have gone so far as to build martin houses among their trees, believing that the toll of insects taken by the birds will materially increase the fruit yield. They are guardians of the barn-yard, since they protect the chickens from the birds of prey. They do not do a single thing that is hurtful to man and many things that are to his benefit. Despite this fact, the modern farmers are tending to discourage the swallows. They leave no openings in their barns through which they can enter. They even break down the nests of the eave swallows on their barns because they mar the smoothness of their walls. They put up no martin nests. All of which is undoubtedly a mistake which means dollar losses, for the presence of these birds is beneficial to farm, village, or city.

QUESTIONS

1. What is the Indian legend of the origin of the swallow?
2. Describe the flight of swallows. How do their feet compare with those of other birds?

3. Compare their bills with those of other birds. Show how the bills are especially made for catching insects. Where is the swallow hunting ground? How do they help the cows?
4. Where do swallows spend the winter? For what do they wait before coming north? Trace on the map the route by which they come.
5. Are swallows kin of the chimney swifts? There are half a dozen families of swallow cousins. Describe the tree swallow. Its manner of life. What change has man brought about in the life of the tree swallow?
6. Long ago where did the cliff swallow build its nest? What change has come about since that time? What is their new name? Tell the story of the sealed nest.
7. Where does the barn swallow build? Tell any experiences that you may have had with swallows. How does the barn swallow differ from other swallows?
8. Describe the homes of bank swallows.
9. Why did the Indians hang up gourds as nests for purple martins? Why do the negroes do the same today? How may the owner of an orchard make use of the martins?
10. Describe the martin houses that have become common. Their contests with English sparrows. Which is master?
11. Where did the martins nest before white men came to America? What may they do now if man does not provide a nest?
12. How many eggs do the swallows lay? What does this show?
13. Are swallows friends or enemies to man? What service do they render?
14. What mistake do some farmers make about the service of the swallow? What should be done to encourage swallows and martins?
15. Contrast swallows and bobwhites. Swallows and wrens. What can you say about the differences in groups of birds?

CHAPTER XXIV

THE CHICKADEE

A "SCRAP of valor," the chickadee has been called because of its defiance, tiny as it is, of so big a thing as a northern winter.

Observe one of them there in the woods of a January day with the snow on the ground and the frozen branches crackling in the wind.

It is the picture of unconcern, of cheerfulness. A cutting blast comes out of the North and the little chickadee scorns to give it even so much notice as to turn its bill toward it, as most birds do, that their feathers will not be parted and the cold penetrate to the skin. Let it blow, says the chickadee. It will pay no attention. Let the thermometer play around zero if it will. The cheerfulness of the chickadee remains undisturbed.

The chickadee is undoubtedly the best loved of the small company of winter birds. Unfortunately it is not to be found all over the United States. The mass of the people know it, however, for it ranges as far west as the Rockies, and as far south as the Potomac, and so covers the part of the United States where most of the people live. And then there is the Carolina chickadee, a little different, that lives all the way to the Gulf of Mexico.

Among the winter birds, the junco, cousin to the sparrow, slate-gray above, white below, is a snowbird that is more widely scattered than the chickadee, but with less personality. Then there is the tufted titmouse, built like a jay, with a very charming topknot, pert and jaunty. With its "peco, peco, peco" call it is noisier than the chickadee, but less friendly. It is a close relative, for the chickadee is also a titmouse, the black-capped titmouse. Then there are white-breasted nuthatches, relatives also, that are called tree mice and devil-downheads, from the way they run about the branches, and hang underneath them. And there is the tiny ruby-crowned kinglet, that leaves its summer home, the north woods, in winter, and, though it has never before seen a human being, is not at all afraid of them. All are members of the hardy group of winter birds.

Black-capped, gray-backed, white underneath, stout-billed is the chickadee. It gets its name from its "chicka-dee-dee" call. It has a hunting note also of "day, day, day," which it keeps going quite busily. It talks all the time and always cheerfully. Then there is the "phœ-be" call which is its love song. All are clear as bells, tinkling, liquid, cheery, highly pleasing.

The chickadees like the man creature and when he appears in the woods they follow him and play about him, instead of scurrying away. They learn in his garden to alight on his finger or shoulder, to peep inquiringly under his hat brim. They are the gentlest of birds, with the possible exception of the chipping sparrow, and the most cheerful, with the possible exception of the house wren.

The chickadee is a symbol of faithfulness. It lives the year around in the same region. It never deceives its human friends, as so many birds do, by changing its coat and colors. In the summer, to be sure, it is not much seen. It is, in fact, then nesting in some deep woods. There are so many leaves and so many bigger birds that it is quite lost. But when wintry blasts drive the others south, the chickadee begins to be noticed. Then there comes a time when it is almost the only thing to lend a touch of life and a note of gladness to a bleak out of doors.

The chickadee gets along quite well, thank you, when all the world is buried in snow and the food sources of most creatures are well covered up. It has a method of harvesting that can be worked in the very dead of winter, a source of food supply, or several sources, that it can depend upon even then.

The chickadees, the titmice, and the nuthatches, all close relatives, get their food supply from insect sources even when there seems to be no evidence of the presence of insects. Many of these insects, however, have left grubs or clusters of eggs tucked away in the crevices of the bark of trees that they may lie there through the winter and hatch out into a new generation of insects when spring-time comes. It is the business of these sharp-billed birds to prowl about the trunks and branches of the trees and feed upon these tiny eggs. An egg the size of a pinpoint would not help much to stay the hunger of a bigger bird, but these tiny ones, whose bodies are no bigger than the end of your thumb, do not require a great bulk of food.

All the members of this family are skilled acrobats and trapeze performers. They have sharp claws which clamp into the bark and make it possible for them to walk on the underside of limbs and swing head down from any twig. This upside-down habit of theirs enables them to get at many parts of the tree that are not explored by other insect-hunting birds and that are therefore likely to have reserve supplies of food.

If the grubs and insect eggs are all gone, however, chickadees still have certain other chances to eat. Almost anywhere it will be found that, though the snow

A CHICKADEE

is deep, the stiff, coarse stalks of the primrose, the black-ribbed stems of the pig-weed, and other such plants will stick out above it. There are still seeds clinging to such winter weeds and grasses. If the stalks are shaken these seeds will probably rattle out and fall down upon the snow-white tablecloth that has been spread below, and they will be easily found.

Failing this supply, the chickadee is likely to find some friend among its human neighbors who has provided for its comfort. The most popular dish that man can supply for the chickadee is suet, and bits of suet on some shelf near the house, made for the purpose, is likely to attract many of these birds. In fact, the chickadee offers an exchange of services which many a wise orchardman is glad to accept. If he will put out these suet stores here and there among his trees, these birds will make headquarters thereabouts. They will eat suet, but between times they will hunt insect eggs. This will mean money to the orchardman when the fruit-picking season comes around the next summer. The little winter visitors will devour eggs that otherwise would have developed into canker worms, plant lice, or many another of the pests of summer. There is a possibility here of helping the summer crop by developing winter bird friendships, which few orchardmen appreciate.

If you have a feeding table for the birds in your window, it is quite easy to establish even closer friendship with the chickadees. Sunflower seed is one of their favorite foods. An excellent way to encourage chickadees is to grow a few sunflower stalks and let them stand. There is much interest in watching this tiny bird husk a sunflower seed before swallowing it. The husk is quite tough, as you will find if you try to take it off with your finger nail. Yet the tiny chickadee gets rid of it with two or three strokes of its sturdy bill.

Go into the garden and hold out your hand with some

of these sunflower seeds in the palm, and if you will be patient about it, you will find the bait gradually approached by your little winter neighbors. They will come quicker if you can whistle their call. Presently there will be a dozen of them in the bare branches of the apple tree over your head. Finally one will perch on the twig that is very near you. It will eye the sunflower seeds quite longingly. Presently it will flit from its twig and almost alight on your finger. Just before it alights, however, fear will overcome it. It will put on its wing brakes, stop, and fly away. After repeating this approach for two or three times, however, it will alight on your finger, get itself a sunflower seed, and dart off in a great fright.

The bright eyes of the other chickadees will have been upon it in this experiment. A second one will follow its example, and then another and another. After a few days the chickadee family is likely to make itself quite at ease in taking sunflower seed from your hand, and if you will be very patient they will come to know you so well that many of them will alight on your fingers, arms, and shoulders whenever you come into the garden.

Yet these chickadees are not to say dependent upon man. Most of them live along the fringes of woods in the winter, quite away from him. Dense pines and cedars are the winter homes of these birds and, should you climb one of these with a lantern, of a winter night, you would be surprised at the numbers of such residents.

It is a marvel, to be sure, that these tiny creatures do not freeze to death in the winter time. The fact is that the rough feathers which are seen on the chickadee are but its overcoat. If the flap of this overcoat blows back the bird is still protected, for there is a layer of fine, downy feathers beneath it, making up a very close, warm garment. Then if you examine the plump little body of the chickadee, you will find that it protects itself from the cold just as does that biggest of animals, the whale. It stores layers of fat for the purpose, just beneath its skin.

CHICKADEE AND HER YOUNG

Despite all of this the chickadee has within it a little furnace that consumes a considerable amount of fuel. It maintains a temperature within itself that is nearly ten degrees higher than that maintained by man, sitting there in his flannels in a heated house. Its food is its fuel and, for its size, it consumes a surprising amount of it. Tests have been made and it has been shown that a single chickadee will eat twenty sunflower seeds in an hour. It stokes the furnace constantly and keeps a busy little fire raging throughout the winter. If it were not for its first-class little heating system this black and white midget would not be able to live where it does through the winter.

The chickadee has certain traits which are like those of the woodpecker. It is quite a good worker with its bill. It drills successfully into a piece of frozen suet. It will attack an ear of corn, drilling into the grains from the top and eating the heart out of them. It digs insect eggs out of crevices in the bark. Finally, it will, upon occasion, dig itself out a nest in the trunk of a tree just as does the woodpecker. To be sure, it prefers to appropriate the old nest of some other bird, or to find a natural cavity in which to build. And if it digs its own nest it usually picks a decayed and quite soft bit of wood. Here it may rear six or eight little ones, families that are large and hearty, guaranteeing generations of this bit of cheerfulness left behind by summer as a pledge when most other life of the out of doors has disappeared.

QUESTIONS

1. In what parts of the United States is the chickadee to be found in winter? How does it act in zero weather?
2. What are some of the other winter birds? To what family does the junco belong? Describe three cousins of the chickadee.
3. What are the three calls of the chickadee? What are its relations to man?
4. Does the chickadee migrate? Where is it not much seen in summer?
5. What insect food do chickadees and their relatives find when there are no insects? Why do they learn to walk upside down?

6. What do they eat when there are no insect eggs? What is the favorite chickadee food that man may supply? How will the chickadee pay the orchardman for suet?
7. How may chickadees be kept about the house? How do they husk a sunflower seed? How can they be taught to light on one's hand?
8. What trees furnish winter roosting places for these hardy little birds? Show how the chickadee protects itself against the cold. Describe its heating plant. The amount of fuel it burns.
9. What feats does the chickadee perform with its bill? Where does it make its nest? How many little ones does it raise?
10. What winter birds live about your home? Tell some of the things that you have seen them do.

INDEX

A CATALOGUE OF SELECTED DOVER BOOKS IN ALL FIELDS OF INTEREST

A CATALOGUE OF SELECTED DOVER BOOKS IN ALL FIELDS OF INTEREST

WHAT IS SCIENCE?, *N. Campbell*

The role of experiment and measurement, the function of mathematics, the nature of scientific laws, the difference between laws and theories, the limitations of science, and many similarly provocative topics are treated clearly and without technicalities by an eminent scientist. "Still an excellent introduction to scientific philosophy," H. Margenau in *Physics Today*. "A first-rate primer . . . deserves a wide audience," *Scientific American*. 192pp. 5⅜ x 8.

S43 Paperbound $1.25

THE NATURE OF LIGHT AND COLOUR IN THE OPEN AIR, *M. Minnaert*

Why are shadows sometimes blue, sometimes green, or other colors depending on the light and surroundings? What causes mirages? Why do multiple suns and moons appear in the sky? Professor Minnaert explains these unusual phenomena and hundreds of others in simple, easy-to-understand terms based on optical laws and the properties of light and color. No mathematics is required but artists, scientists, students, and everyone fascinated by these "tricks" of nature will find thousands of useful and amazing pieces of information. Hundreds of observational experiments are suggested which require no special equipment. 200 illustrations; 42 photos. xvi + 362pp. 5⅜ x 8.

T196 Paperbound $2.00

THE STRANGE STORY OF THE QUANTUM, AN ACCOUNT FOR THE GENERAL READER OF THE GROWTH OF IDEAS UNDERLYING OUR PRESENT ATOMIC KNOWLEDGE, *B. Hoffmann*

Presents lucidly and expertly, with barest amount of mathematics, the problems and theories which led to modern quantum physics. Dr. Hoffmann begins with the closing years of the 19th century, when certain trifling discrepancies were noticed, and with illuminating analogies and examples takes you through the brilliant concepts of Planck, Einstein, Pauli, Broglie, Bohr, Schroedinger, Heisenberg, Dirac, Sommerfeld, Feynman, etc. This edition includes a new, long postscript carrying the story through 1958. "Of the books attempting an account of the history and contents of our modern atomic physics which have come to my attention, this is the best," H. Margenau, Yale University, in *American Journal of Physics*. 32 tables and line illustrations. Index. 275pp. 5⅜ x 8. T518 Paperbound $2.00

GREAT IDEAS OF MODERN MATHEMATICS: THEIR NATURE AND USE, *Jagjit Singh*

Reader with only high school math will understand main mathematical ideas of modern physics, astronomy, genetics, psychology, evolution, etc. better than many who use them as tools, but comprehend little of their basic structure. Author uses his wide knowledge of non-mathematical fields in brilliant exposition of differential equations, matrices, group theory, logic, statistics, problems of mathematical foundations, imaginary numbers, vectors, etc. Original publication. 2 appendixes. 2 indexes. 65 ills. 322pp. 5⅜ x 8.

T587 Paperbound $2.00

THE MUSIC OF THE SPHERES: THE MATERIAL UNIVERSE—FROM ATOM TO QUASAR, SIMPLY EXPLAINED, *Guy Murchie*
Vast compendium of fact, modern concept and theory, observed and calculated data, historical background guides intelligent layman through the material universe. Brilliant exposition of earth's construction, explanations for moon's craters, atmospheric components of Venus and Mars (with data from recent fly-by's), sun spots, sequences of star birth and death, neighboring galaxies, contributions of Galileo, Tycho Brahe, Kepler, etc.; and (Vol. 2) construction of the atom (describing newly discovered sigma and xi subatomic particles), theories of sound, color and light, space and time, including relativity theory, quantum theory, wave theory, probability theory, work of Newton, Maxwell, Faraday, Einstein, de Broglie, etc. "Best presentation yet offered to the intelligent general reader," *Saturday Review*. Revised (1967). Index. 319 illustrations by the author. Total of xx + 644pp. 5⅜ x 8½.
T1809, T1810 Two volume set, paperbound $4.00

FOUR LECTURES ON RELATIVITY AND SPACE, *Charles Proteus Steinmetz*
Lecture series, given by great mathematician and electrical engineer, generally considered one of the best popular-level expositions of special and general relativity theories and related questions. Steinmetz translates complex mathematical reasoning into language accessible to laymen through analogy, example and comparison. Among topics covered are relativity of motion, location, time; of mass; acceleration; 4-dimensional time-space; geometry of the gravitational field; curvature and bending of space; non-Euclidean geometry. Index. 40 illustrations. x + 142pp. 5⅜ x 8½. S1771 Paperbound $1.35

HOW TO KNOW THE WILD FLOWERS, *Mrs. William Starr Dana*
Classic nature book that has introduced thousands to wonders of American wild flowers. Color-season principle of organization is easy to use, even by those with no botanical training, and the genial, refreshing discussions of history, folklore, uses of over 1,000 native and escape flowers, foliage plants are informative as well as fun to read. Over 170 full-page plates, collected from several editions, may be colored in to make permanent records of finds. Revised to conform with 1950 edition of Gray's Manual of Botany. xlii + 438pp. 5⅜ x 8½. T332 Paperbound $2.25

MANUAL OF THE TREES OF NORTH AMERICA, *Charles Sprague Sargent*
Still unsurpassed as most comprehensive, reliable study of North American tree characteristics, precise locations and distribution. By dean of American dendrologists. Every tree native to U.S., Canada, Alaska; 185 genera, 717 species, described in detail—leaves, flowers, fruit, winterbuds, bark, wood, growth habits, etc. plus discussion of varieties and local variants, immaturity variations. Over 100 keys, including unusual 11-page analytical key to genera, aid in identification. 783 clear illustrations of flowers, fruit, leaves. An unmatched permanent reference work for all nature lovers. Second enlarged (1926) edition. Synopsis of families. Analytical key to genera. Glossary of technical terms. Index. 783 illustrations, 1 map. Total of 982pp. 5⅜ x 8.
T277, T278 Two volume set, paperbound $5.00

IT'S FUN TO MAKE THINGS FROM SCRAP MATERIALS, *Evelyn Glantz Hershoff*
What use are empty spools, tin cans, bottle tops? What can be made from rubber bands, clothes pins, paper clips, and buttons? This book provides simply worded instructions and large diagrams showing you how to make cookie cutters, toy trucks, paper turkeys, Halloween masks, telephone sets, aprons, linoleum block- and spatter prints — in all 399 projects! Many are easy enough for young children to figure out for themselves; some challenging enough to entertain adults; all are remarkably ingenious ways to make things from materials that cost pennies or less! Formerly "Scrap Fun for Everyone." Index. 214 illustrations. 373pp. 5⅜ x 8½. T1251 Paperbound $1.50

SYMBOLIC LOGIC and THE GAME OF LOGIC, *Lewis Carroll*
"Symbolic Logic" is not concerned with modern symbolic logic, but is instead a collection of over 380 problems posed with charm and imagination, using the syllogism and a fascinating diagrammatic method of drawing conclusions. In "The Game of Logic" Carroll's whimsical imagination devises a logical game played with 2 diagrams and counters (included) to manipulate hundreds of tricky syllogisms. The final section, "Hit or Miss" is a lagniappe of 101 additional puzzles in the delightful Carroll manner. Until this reprint edition, both of these books were rarities costing up to $15 each. Symbolic Logic: Index. xxxi + 199pp. The Game of Logic: 96pp. 2 vols. bound as one. 5⅜ x 8. T492 Paperbound $2.00

MATHEMATICAL PUZZLES OF SAM LOYD, PART I
selected and edited by M. Gardner
Choice puzzles by the greatest American puzzle creator and innovator. Selected from his famous collection, "Cyclopedia of Puzzles," they retain the unique style and historical flavor of the originals. There are posers based on arithmetic, algebra, probability, game theory, route tracing, topology, counter and sliding block, operations research, geometrical dissection. Includes the famous "14-15" puzzle which was a national craze, and his "Horse of a Different Color" which sold millions of copies. 117 of his most ingenious puzzles in all. 120 line drawings and diagrams. Solutions. Selected references. xx + 167pp. 5⅜ x 8. T498 Paperbound $1.25

STRING FIGURES AND HOW TO MAKE THEM, *Caroline Furness Jayne*
107 string figures plus variations selected from the best primitive and modern examples developed by Navajo, Apache, pygmies of Africa, Eskimo, in Europe, Australia, China, etc. The most readily understandable, easy-to-follow book in English on perennially popular recreation. Crystal-clear exposition; step-by-step diagrams. Everyone from kindergarten children to adults looking for unusual diversion will be endlessly amused. Index. Bibliography. Introduction by A. C. Haddon. 17 full-page plates, 960 illustrations. xxiii + 401pp. 5⅜ x 8½. T152 Paperbound $2.25

PAPER FOLDING FOR BEGINNERS, *W. D. Murray and F. J. Rigney*
A delightful introduction to the varied and entertaining Japanese art of origami (paper folding), with a full, crystal-clear text that anticipates every difficulty; over 275 clearly labeled diagrams of all important stages in creation. You get results at each stage, since complex figures are logically developed from simpler ones. 43 different pieces are explained: sailboats, frogs, roosters, etc. 6 photographic plates. 279 diagrams. 95pp. 5⅝ x 8⅜. T713 Paperbound $1.00

PRINCIPLES OF ART HISTORY,
H. Wölfflin

Analyzing such terms as "baroque," "classic," "neoclassic," "primitive," "picturesque," and 164 different works by artists like Botticelli, van Cleve, Dürer, Hobbema, Holbein, Hals, Rembrandt, Titian, Brueghel, Vermeer, and many others, the author establishes the classifications of art history and style on a firm, concrete basis. This classic of art criticism shows what really occurred between the 14th-century primitives and the sophistication of the 18th century in terms of basic attitudes and philosophies. "A remarkable lesson in the art of seeing," *Sat. Rev. of Literature.* Translated from the 7th German edition. 150 illustrations. 254pp. 6⅛ x 9¼. T276 Paperbound $2.00

PRIMITIVE ART,
Franz Boas

This authoritative and exhaustive work by a great American anthropologist covers the entire gamut of primitive art. Pottery, leatherwork, metal work, stone work, wood, basketry, are treated in detail. Theories of primitive art, historical depth in art history, technical virtuosity, unconscious levels of patterning, symbolism, styles, literature, music, dance, etc. A must book for the interested layman, the anthropologist, artist, handicrafter (hundreds of unusual motifs), and the historian. Over 900 illustrations (50 ceramic vessels, 12 totem poles, etc.). 376pp. 5⅜ x 8. T25 Paperbound $2.50

THE GENTLEMAN AND CABINET MAKER'S DIRECTOR,
Thomas Chippendale

A reprint of the 1762 catalogue of furniture designs that went on to influence generations of English and Colonial and Early Republic American furniture makers. The 200 plates, most of them full-page sized, show Chippendale's designs for French (Louis XV), Gothic, and Chinese-manner chairs, sofas, canopy and dome beds, cornices, chamber organs, cabinets, shaving tables, commodes, picture frames, frets, candle stands, chimney pieces, decorations, etc. The drawings are all elegant and highly detailed; many include construction diagrams and elevations. A supplement of 24 photographs shows surviving pieces of original and Chippendale-style pieces of furniture. Brief biography of Chippendale by N. I. Bienenstock, editor of *Furniture World.* Reproduced from the 1762 edition. 200 plates, plus 19 photographic plates. vi + 249pp. 9⅛ x 12¼. T1601 Paperbound $3.50

AMERICAN ANTIQUE FURNITURE: A BOOK FOR AMATEURS,
Edgar G. Miller, Jr.

Standard introduction and practical guide to identification of valuable American antique furniture. 2115 illustrations, mostly photographs taken by the author in 148 private homes, are arranged in chronological order in extensive chapters on chairs, sofas, chests, desks, bedsteads, mirrors, tables, clocks, and other articles. Focus is on furniture accessible to the collector, including simpler pieces and a larger than usual coverage of Empire style. Introductory chapters identify structural elements, characteristics of various styles, how to avoid fakes, etc. "We are frequently asked to name some book on American furniture that will meet the requirements of the novice collector, the beginning dealer, and . . . the general public. . . . We believe Mr. Miller's two volumes more completely satisfy this specification than any other work," *Antiques.* Appendix. Index. Total of vi + 1106pp. 7⅞ x 10¾.
T1599, T1600 Two volume set, paperbound $7.50

THE BAD CHILD'S BOOK OF BEASTS, MORE BEASTS FOR WORSE CHILDREN, and A MORAL ALPHABET, *H. Belloc*
Hardly and anthology of humorous verse has appeared in the last 50 years without at least a couple of these famous nonsense verses. But one must see the entire volumes—with all the delightful original illustrations by Sir Basil Blackwood—to appreciate fully Belloc's charming and witty verses that play so subacidly on the platitudes of life and morals that beset his day—and ours. A great humor classic. Three books in one. Total of 157pp. 5⅜ x 8.
T749 Paperbound $1.00

THE DEVIL'S DICTIONARY, *Ambrose Bierce*
Sardonic and irreverent barbs puncturing the pomposities and absurdities of American politics, business, religion, literature, and arts, by the country's greatest satirist in the classic tradition. Epigrammatic as Shaw, piercing as Swift, American as Mark Twain, Will Rogers, and Fred Allen, Bierce will always remain the favorite of a small coterie of enthusiasts, and of writers and speakers whom he supplies with "some of the most gorgeous witticisms of the English language" (H. L. Mencken). Over 1000 entries in alphabetical order. 144pp. 5⅜ x 8. T487 Paperbound $1.00

THE COMPLETE NONSENSE OF EDWARD LEAR.
This is the only complete edition of this master of gentle madness available at a popular price. *A Book of Nonsense, Nonsense Songs, More Nonsense Songs and Stories* in their entirety with all the old favorites that have delighted children and adults for years. The Dong With A Luminous Nose, The Jumblies, The Owl and the Pussycat, and hundreds of other bits of wonderful nonsense. 214 limericks, 3 sets of Nonsense Botany, 5 Nonsense Alphabets, 546 drawings by Lear himself, and much more. 320pp. 5⅜ x 8. T167 Paperbound $1.50

THE WIT AND HUMOR OF OSCAR WILDE, *ed. by Alvin Redman*
Wilde at his most brilliant, in 1000 epigrams exposing weaknesses and hypocrisies of "civilized" society. Divided into 49 categories—sin, wealth, women, America, etc.—to aid writers, speakers. Includes excerpts from his trials, books, plays, criticism. Formerly "The Epigrams of Oscar Wilde." Introduction by Vyvyan Holland, Wilde's only living son. Introductory essay by editor. 260pp. 5⅜ x 8. T602 Paperbound $1.50

A CHILD'S PRIMER OF NATURAL HISTORY, *Oliver Herford*
Scarcely an anthology of whimsy and humor has appeared in the last 50 years without a contribution from Oliver Herford. Yet the works from which these examples are drawn have been almost impossible to obtain! Here at last are Herford's improbable definitions of a menagerie of familiar and weird animals, each verse illustrated by the author's own drawings. 24 drawings in 2 colors; 24 additional drawings. vii + 95pp. 6½ x 6. T1647 Paperbound $1.00

THE BROWNIES: THEIR BOOK, *Palmer Cox*
The book that made the Brownies a household word. Generations of readers have enjoyed the antics, predicaments and adventures of these jovial sprites, who emerge from the forest at night to play or to come to the aid of a deserving human. Delightful illustrations by the author decorate nearly every page. 24 short verse tales with 266 illustrations. 155pp. 6⅝ x 9¼.
T1265 Paperbound $1.50

THE PRINCIPLES OF PSYCHOLOGY,
William James

The full long-course, unabridged, of one of the great classics of Western literature and science. Wonderfully lucid descriptions of human mental activity, the stream of thought, consciousness, time perception, memory, imagination, emotions, reason, abnormal phenomena, and similar topics. Original contributions are integrated with the work of such men as Berkeley, Binet, Mills, Darwin, Hume, Kant, Royce, Schopenhauer, Spinoza, Locke, Descartes, Galton, Wundt, Lotze, Herbart, Fechner, and scores of others. All contrasting interpretations of mental phenomena are examined in detail—introspective analysis, philosophical interpretation, and experimental research. "A classic," *Journal of Consulting Psychology*. "The main lines are as valid as ever," *Psychoanalytical Quarterly*. "Standard reading . . . a classic of interpretation," *Psychiatric Quarterly*. 94 illustrations. 1408pp. 5⅜ x 8.
T381, T382 Two volume set, paperbound $5.25

VISUAL ILLUSIONS: THEIR CAUSES, CHARACTERISTICS AND APPLICATIONS,
M. Luckiesh

"Seeing is deceiving," asserts the author of this introduction to virtually every type of optical illusion known. The text both describes and explains the principles involved in color illusions, figure-ground, distance illusions, etc. 100 photographs, drawings and diagrams prove how easy it is to fool the sense: circles that aren't round, parallel lines that seem to bend, stationary figures that seem to move as you stare at them – illustration after illustration strains our credulity at what we see. Fascinating book from many points of view, from applications for artists, in camouflage, etc. to the psychology of vision. New introduction by William Ittleson, Dept. of Psychology, Queens College. Index. Bibliography. xxi + 252pp. 5⅜ x 8½. T1530 Paperbound $1.50

FADS AND FALLACIES IN THE NAME OF SCIENCE,
Martin Gardner

This is the standard account of various cults, quack systems, and delusions which have masqueraded as science: hollow earth fanatics. Reich and orgone sex energy, dianetics, Atlantis, multiple moons, Forteanism, flying saucers, medical fallacies like iridiagnosis, zone therapy, etc. A new chapter has been added on Bridey Murphy, psionics, and other recent manifestations in this field. This is a fair, reasoned appraisal of eccentric theory which provides excellent inoculation against cleverly masked nonsense. "Should be read by everyone, scientist and non-scientist alike," R. T. Birge, Prof. Emeritus of Physics, Univ. of California; Former President, American Physical Society. Index. x + 365pp. 5⅜ x 8. T394 Paperbound $2.00

ILLUSIONS AND DELUSIONS OF THE SUPERNATURAL AND THE OCCULT,
D. H. Rawcliffe

Holds up to rational examination hundreds of persistent delusions including crystal gazing, automatic writing, table turning, mediumistic trances, mental healing, stigmata, lycanthropy, live burial, the Indian Rope Trick, spiritualism, dowsing, telepathy, clairvoyance, ghosts, ESP, etc. The author explains and exposes the mental and physical deceptions involved, making this not only an exposé of supernatural phenomena, but a valuable exposition of characteristic types of abnormal psychology. Originally titled "The Psychology of the Occult." 14 illustrations. Index. 551pp. 5⅜ x 8. T503 Paperbound $2.75

FAIRY TALE COLLECTIONS, *edited by Andrew Lang*
Andrew Lang's fairy tale collections make up the richest shelf-full of traditional children's stories anywhere available. Lang supervised the translation of stories from all over the world—familiar European tales collected by Grimm, animal stories from Negro Africa, myths of primitive Australia, stories from Russia, Hungary, Iceland, Japan, and many other countries. Lang's selection of translations are unusually high; many authorities consider that the most familiar tales find their best versions in these volumes. All collections are richly decorated and illustrated by H. J. Ford and other artists.

THE BLUE FAIRY BOOK. 37 stories. 138 illustrations. ix + 390pp. 5⅜ x 8½.
T1437 Paperbound $1.75

THE GREEN FAIRY BOOK. 42 stories. 100 illustrations. xiii + 366pp. 5⅜ x 8½.
T1439 Paperbound $1.75

THE BROWN FAIRY BOOK. 32 stories. 50 illustrations, 8 in color. xii + 350pp. 5⅜ x 8½.
T1438 Paperbound $1.95

THE BEST TALES OF HOFFMANN, *edited by E. F. Bleiler*
10 stories by E. T. A. Hoffmann, one of the greatest of all writers of fantasy. The tales include "The Golden Flower Pot," "Automata," "A New Year's Eve Adventure," "Nutcracker and the King of Mice," "Sand-Man," and others. Vigorous characterizations of highly eccentric personalities, remarkably imaginative situations, and intensely fast pacing has made these tales popular all over the world for 150 years. Editor's introduction. 7 drawings by Hoffmann. xxxiii + 419pp. 5⅜ x 8½.
T1793 Paperbound $2.25

GHOST AND HORROR STORIES OF AMBROSE BIERCE,
edited by E. F. Bleiler
Morbid, eerie, horrifying tales of possessed poets, shabby aristocrats, revived corpses, and haunted malefactors. Widely acknowledged as the best of their kind between Poe and the moderns, reflecting their author's inner torment and bitter view of life. Includes "Damned Thing," "The Middle Toe of the Right Foot," "The Eyes of the Panther," "Visions of the Night," "Moxon's Master," and over a dozen others. Editor's introduction. xxii + 199pp. 5⅜ x 8½.
T767 Paperbound $1.50

THREE GOTHIC NOVELS, *edited by E. F. Bleiler*
Originators of the still popular Gothic novel form, influential in ushering in early 19th-century Romanticism. Horace Walpole's *Castle of Otranto*, William Beckford's *Vathek*, John Polidori's *The Vampyre*, and a *Fragment* by Lord Byron are enjoyable as exciting reading or as documents in the history of English literature. Editor's introduction. xi + 291pp. 5⅜ x 8½.
T1232 Paperbound $2.00

BEST GHOST STORIES OF LEFANU, *edited by E. F. Bleiler*
Though admired by such critics as V. S. Pritchett, Charles Dickens and Henry James, ghost stories by the Irish novelist Joseph Sheridan LeFanu have never become as widely known as his detective fiction. About half of the 16 stories in this collection have never before been available in America. Collection includes "Carmilla" (perhaps the best vampire story ever written), "The Haunted Baronet," "The Fortunes of Sir Robert Ardagh," and the classic "Green Tea." Editor's introduction. 7 contemporary illustrations. Portrait of LeFanu. xii + 467pp. 5⅜ x 8.
T415 Paperbound $2.00

EASY-TO-DO ENTERTAINMENTS AND DIVERSIONS WITH COINS, CARDS, STRING, PAPER AND MATCHES, *R. M. Abraham*

Over 300 tricks, games and puzzles will provide young readers with absorbing fun. Sections on card games; paper-folding; tricks with coins, matches and pieces of string; games for the agile; toy-making from common household objects; mathematical recreations; and 50 miscellaneous pastimes. Anyone in charge of groups of youngsters, including hard-pressed parents, and in need of suggestions on how to keep children sensibly amused and quietly content will find this book indispensable. Clear, simple text, copious number of delightful line drawings and illustrative diagrams. Originally titled "Winter Nights' Entertainments." Introduction by Lord Baden Powell. 329 illustrations. v + 186pp. 5⅜ x 8½. T921 Paperbound $1.00

AN INTRODUCTION TO CHESS MOVES AND TACTICS SIMPLY EXPLAINED, *Leonard Barden*

Beginner's introduction to the royal game. Names, possible moves of the pieces, definitions of essential terms, how games are won, etc. explained in 30-odd pages. With this background you'll be able to sit right down and play. Balance of book teaches strategy – openings, middle game, typical endgame play, and suggestions for improving your game. A sample game is fully analyzed. True middle-level introduction, teaching you all the essentials without oversimplifying or losing you in a maze of detail. 58 figures. 102pp. 5⅜ x 8½. T1210 Paperbound $1.25

LASKER'S MANUAL OF CHESS, *Dr. Emanuel Lasker*

Probably the greatest chess player of modern times, Dr. Emanuel Lasker held the world championship 28 years, independent of passing schools or fashions. This unmatched study of the game, chiefly for intermediate to skilled players, analyzes basic methods, combinations, position play, the aesthetics of chess, dozens of different openings, etc., with constant reference to great modern games. Contains a brilliant exposition of Steinitz's important theories. Introduction by Fred Reinfeld. Tables of Lasker's tournament record. 3 indices. 308 diagrams. 1 photograph. xxx + 349pp. 5⅜ x 8. T640 Paperbound $2.25

COMBINATIONS: THE HEART OF CHESS, *Irving Chernev*

Step-by-step from simple combinations to complex, this book, by a well-known chess writer, shows you the intricacies of pins, counter-pins, knight forks, and smothered mates. Other chapters show alternate lines of play to those taken in actual championship games; boomerang combinations; classic examples of brilliant combination play by Nimzovich, Rubinstein, Tarrasch, Botvinnik, Alekhine and Capablanca. Index. 356 diagrams. ix + 245pp. 5⅜ x 8½. T1744 Paperbound $2.00

HOW TO SOLVE CHESS PROBLEMS, *K. S. Howard*

Full of practical suggestions for the fan or the beginner – who knows only the moves of the chessmen. Contains preliminary section and 58 two-move, 46 three-move, and 8 four-move problems composed by 27 outstanding American problem creators in the last 30 years. Explanation of all terms and exhaustive index. "Just what is wanted for the student," Brian Harley. 112 problems, solutions. vi + 171pp. 5⅜ x 8. T748 Paperbound $1.35

SOCIAL THOUGHT FROM LORE TO SCIENCE,
H. E. Barnes and H. Becker
An immense survey of sociological thought and ways of viewing, studying, planning, and reforming society from earliest times to the present. Includes thought on society of preliterate peoples, ancient non-Western cultures, and every great movement in Europe, America, and modern Japan. Analyzes hundreds of great thinkers: Plato, Augustine, Bodin, Vico, Montesquieu, Herder, Comte, Marx, etc. Weighs the contributions of utopians, sophists, fascists and communists; economists, jurists, philosophers, ecclesiastics, and every 19th and 20th century school of scientific sociology, anthropology, and social psychology throughout the world. Combines topical, chronological, and regional approaches, treating the evolution of social thought as a process rather than as a series of mere topics. "Impressive accuracy, competence, and discrimination . . . easily the best single survey," *Nation.* Thoroughly revised, with new material up to 1960. 2 indexes. Over 2200 bibliographical notes. Three volume set. Total of 1586pp. 5⅜ x 8.
T901, T902, T903 Three volume set, paperbound $8.50

A HISTORY OF HISTORICAL WRITING, *Harry Elmer Barnes*
Virtually the only adequate survey of the whole course of historical writing in a single volume. Surveys developments from the beginnings of historiography in the ancient Near East and the Classical World, up through the Cold War. Covers major historians in detail, shows interrelationship with cultural background, makes clear individual contributions, evaluates and estimates importance; also enormously rich upon minor authors and thinkers who are usually passed over. Packed with scholarship and learning, clear, easily written. Indispensable to every student of history. Revised and enlarged up to 1961. Index and bibliography. xv + 442pp. 5⅜ x 8½.
T104 Paperbound $2.50

JOHANN SEBASTIAN BACH, *Philipp Spitta*
The complete and unabridged text of the definitive study of Bach. Written some 70 years ago, it is still unsurpassed for its coverage of nearly all aspects of Bach's life and work. There could hardly be a finer non-technical introduction to Bach's music than the detailed, lucid analyses which Spitta provides for hundreds of individual pieces. 26 solid pages are devoted to the B minor mass, for example, and 30 pages to the glorious St. Matthew Passion. This monumental set also includes a major analysis of the music of the 18th century: Buxtehude, Pachelbel, etc. "Unchallenged as the last word on one of the supreme geniuses of music," John Barkham, *Saturday Review Syndicate.* Total of 1819pp. Heavy cloth binding. 5⅜ x 8.
T252 Two volume set, clothbound $15.00

BEETHOVEN AND HIS NINE SYMPHONIES, *George Grove*
In this modern middle-level classic of musicology Grove not only analyzes all nine of Beethoven's symphonies very thoroughly in terms of their musical structure, but also discusses the circumstances under which they were written, Beethoven's stylistic development, and much other background material. This is an extremely rich book, yet very easily followed; it is highly recommended to anyone seriously interested in music. Over 250 musical passages. Index. viii + 407pp. 5⅜ x 8. T334 Paperbound $2.25

THREE SCIENCE FICTION NOVELS,
John Taine
Acknowledged by many as the best SF writer of the 1920's, Taine (under the name Eric Temple Bell) was also a Professor of Mathematics of considerable renown. Reprinted here are *The Time Stream,* generally considered Taine's best, *The Greatest Game,* a biological-fiction novel, and *The Purple Sapphire,* involving a supercivilization of the past. Taine's stories tie fantastic narratives to frameworks of original and logical scientific concepts. Speculation is often profound on such questions as the nature of time, concept of entropy, cyclical universes, etc. 4 contemporary illustrations. v + 532pp. 5⅜ x 8⅜.
T1180 Paperbound $2.00

SEVEN SCIENCE FICTION NOVELS,
H. G. Wells
Full unabridged texts of 7 science-fiction novels of the master. Ranging from biology, physics, chemistry, astronomy, to sociology and other studies, Mr. Wells extrapolates whole worlds of strange and intriguing character. "One will have to go far to match this for entertainment, excitement, and sheer pleasure . . ."*New York Times.* Contents: The Time Machine, The Island of Dr. Moreau, The First Men in the Moon, The Invisible Man, The War of the Worlds, The Food of the Gods, In The Days of the Comet. 1015pp. 5⅜ x 8.
T264 Clothbound $5.00

28 SCIENCE FICTION STORIES OF H. G. WELLS.
Two full, unabridged novels, *Men Like Gods* and *Star Begotten,* plus 26 short stories by the master science-fiction writer of all time! Stories of space, time, invention, exploration, futuristic adventure. Partial contents: *The Country of the Blind, In the Abyss, The Crystal Egg, The Man Who Could Work Miracles, A Story of Days to Come, The Empire of the Ants, The Magic Shop, The Valley of the Spiders, A Story of the Stone Age, Under the Knife, Sea Raiders,* etc. An indispensable collection for the library of anyone interested in science fiction adventure. 928pp. 5⅜ x 8. T265 Clothbound $5.00

THREE MARTIAN NOVELS,
Edgar Rice Burroughs
Complete, unabridged reprinting, in one volume, of Thuvia, Maid of Mars; Chessmen of Mars; The Master Mind of Mars. Hours of science-fiction adventure by a modern master storyteller. Reset in large clear type for easy reading. 16 illustrations by J. Allen St. John. vi + 490pp. 5⅜ x 8½.
T39 Paperbound $2.50

AN INTELLECTUAL AND CULTURAL HISTORY OF THE WESTERN WORLD,
Harry Elmer Barnes
Monumental 3-volume survey of intellectual development of Europe from primitive cultures to the present day. Every significant product of human intellect traced through history: art, literature, mathematics, physical sciences, medicine, music, technology, social sciences, religions, jurisprudence, education, etc. Presentation is lucid and specific, analyzing in detail specific discoveries, theories, literary works, and so on. Revised (1965) by recognized scholars in specialized fields under the direction of Prof. Barnes. Revised bibliography. Indexes. 24 illustrations. Total of xxix + 1318pp.
T1275, T1276, T1277 Three volume set, paperbound $7.50

HEAR ME TALKIN' TO YA, *edited by Nat Shapiro and Nat Hentoff*
In their own words, Louis Armstrong, King Oliver, Fletcher Henderson, Bunk Johnson, Bix Beiderbecke, Billy Holiday, Fats Waller, Jelly Roll Morton, Duke Ellington, and many others comment on the origins of jazz in New Orleans and its growth in Chicago's South Side, Kansas City's jam sessions, Depression Harlem, and the modernism of the West Coast schools. Taken from taped conversations, letters, magazine articles, other first-hand sources. Editors' introduction. xvi + 429pp. 5⅜ x 8½. T1726 Paperbound $2.00

THE JOURNAL OF HENRY D. THOREAU
A 25-year record by the great American observer and critic, as complete a record of a great man's inner life as is anywhere available. Thoreau's Journals served him as raw material for his formal pieces, as a place where he could develop his ideas, as an outlet for his interests in wild life and plants, in writing as an art, in classics of literature, Walt Whitman and other contemporaries, in politics, slavery, individual's relation to the State, etc. The Journals present a portrait of a remarkable man, and are an observant social history. Unabridged republication of 1906 edition, Bradford Torrey and Francis H. Allen, editors. Illustrations. Total of 1888pp. 8⅜ x 12¼.
T312, T313 Two volume set, clothbound $25.00

A SHAKESPEARIAN GRAMMAR, *E. A. Abbott*
Basic reference to Shakespeare and his contemporaries, explaining through thousands of quotations from Shakespeare, Jonson, Beaumont and Fletcher, North's *Plutarch* and other sources the grammatical usage differing from the modern. First published in 1870 and written by a scholar who spent much of his life isolating principles of Elizabethan language, the book is unlikely ever to be superseded. Indexes. xxiv + 511pp. 5⅜ x 8½. T1582 Paperbound $2.75

FOLK-LORE OF SHAKESPEARE, *T. F. Thistelton Dyer*
Classic study, drawing from Shakespeare a large body of references to supernatural beliefs, terminology of falconry and hunting, games and sports, good luck charms, marriage customs, folk medicines, superstitions about plants, animals, birds, argot of the underworld, sexual slang of London, proverbs, drinking customs, weather lore, and much else. From full compilation comes a mirror of the 17th-century popular mind. Index. ix + 526pp. 5⅜ x 8½.
T1614 Paperbound $2.75

THE NEW VARIORUM SHAKESPEARE, *edited by H. H. Furness*
By far the richest editions of the plays ever produced in any country or language. Each volume contains complete text (usually First Folio) of the play, all variants in Quarto and other Folio texts, editorial changes by every major editor to Furness's own time (1900), footnotes to obscure references or language, extensive quotes from literature of Shakespearian criticism, essays on plot sources (often reprinting sources in full), and much more.

HAMLET, *edited by H. H. Furness*
Total of xxvi + 905pp. 5⅜ x 8½.
T1004, T1005 Two volume set, paperbound $5.25

TWELFTH NIGHT, *edited by H. H. Furness*
Index. xxii + 434pp. 5⅜ x 8½. T1189 Paperbound $2.75

LA BOHEME BY GIACOMO PUCCINI,
translated and introduced by Ellen H. Bleiler
Complete handbook for the operagoer, with everything needed for full enjoyment except the musical score itself. Complete Italian libretto, with new, modern English line-by-line translation—the only libretto printing all repeats; biography of Puccini; the librettists; background to the opera, Murger's La Boheme, etc.; circumstances of composition and performances; plot summary; and pictorial section of 73 illustrations showing Puccini, famous singers and performances, etc. Large clear type for easy reading. 124pp. 5⅜ x 8½.
T404 Paperbound $1.00

ANTONIO STRADIVARI: HIS LIFE AND WORK (1644-1737),
W. Henry Hill, Arthur F. Hill, and Alfred E. Hill
Still the only book that really delves into life and art of the incomparable Italian craftsman, maker of the finest musical instruments in the world today. The authors, expert violin-makers themselves, discuss Stradivari's ancestry, his construction and finishing techniques, distinguished characteristics of many of his instruments and their locations. Included, too, is story of introduction of his instruments into France, England, first revelation of their supreme merit, and information on his labels, number of instruments made, prices, mystery of ingredients of his varnish, tone of pre-1684 Stradivari violin and changes between 1684 and 1690. An extremely interesting, informative account for all music lovers, from craftsman to concert-goer. Republication of original (1902) edition. New introduction by Sydney Beck, Head of Rare Book and Manuscript Collections, Music Division, New York Public Library. Analytical index by Rembert Wurlitzer. Appendixes. 68 illustrations. 30 full-page plates. 4 in color. xxvi + 315pp. 5⅜ x 8½. T425 Paperbound $2.25

MUSICAL AUTOGRAPHS FROM MONTEVERDI TO HINDEMITH,
Emanuel Winternitz
For beauty, for intrinsic interest, for perspective on the composer's personality, for subtleties of phrasing, shading, emphasis indicated in the autograph but suppressed in the printed score, the mss. of musical composition are fascinating documents which repay close study in many different ways. This 2-volume work reprints facsimiles of mss. by virtually every major composer, and many minor figures—196 examples in all. A full text points out what can be learned from mss., analyzes each sample. Index. Bibliography. 18 figures. 196 plates. Total of 170pp. of text. 7⅞ x 10¾.
T1312, T1313 Two volume set, paperbound $4.00

J. S. BACH,
Albert Schweitzer
One of the few great full-length studies of Bach's life and work, and the study upon which Schweitzer's renown as a musicologist rests. On first appearance (1911), revolutionized Bach performance. The only writer on Bach to be musicologist, performing musician, and student of history, theology and philosophy, Schweitzer contributes particularly full sections on history of German Protestant church music, theories on motivic pictorial representations in vocal music, and practical suggestions for performance. Translated by Ernest Newman. Indexes. 5 illustrations. 650 musical examples. Total of xix + 928pp. 5⅜ x 8½. T1631, T1632 Two volume set, paperbound $4.50

THE METHODS OF ETHICS, *Henry Sidgwick*
Propounding no organized system of its own, study subjects every major methodological approach to ethics to rigorous, objective analysis. Study discusses and relates ethical thought of Plato, Aristotle, Bentham, Clarke, Butler, Hobbes, Hume, Mill, Spencer, Kant, and dozens of others. Sidgwick retains conclusions from each system which follow from ethical premises, rejecting the faulty. Considered by many in the field to be among the most important treatises on ethical philosophy. Appendix. Index. xlvii + 528pp. 5⅜ x 8½.
T1608 Paperbound $2.50

TEUTONIC MYTHOLOGY, *Jakob Grimm*
A milestone in Western culture; the work which established on a modern basis the study of history of religions and comparative religions. 4-volume work assembles and interprets everything available on religious and folkloristic beliefs of Germanic people (including Scandinavians, Anglo-Saxons, etc.). Assembling material from such sources as Tacitus, surviving Old Norse and Icelandic texts, archeological remains, folktales, surviving superstitions, comparative traditions, linguistic analysis, etc. Grimm explores pagan deities, heroes, folklore of nature, religious practices, and every other area of pagan German belief. To this day, the unrivaled, definitive, exhaustive study. Translated by J. S. Stallybrass from 4th (1883) German edition. Indexes. Total of lxxvii + 1887pp. 5⅜ x 8½.
T1602, T1603, T1604, T1605 Four volume set, paperbound $10.00

THE I CHING, *translated by James Legge*
Called "The Book of Changes" in English, this is one of the Five Classics edited by Confucius, basic and central to Chinese thought. Explains perhaps the most complex system of divination known, founded on the theory that all things happening at any one time have characteristic features which can be isolated and related. Significant in Oriental studies, in history of religions and philosophy, and also to Jungian psychoanalysis and other areas of modern European thought. Index. Appendixes. 6 plates. xxi + 448pp. 5⅜ x 8½.
T1062 Paperbound $2.75

HISTORY OF ANCIENT PHILOSOPHY, *W. Windelband*
One of the clearest, most accurate comprehensive surveys of Greek and Roman philosophy. Discusses ancient philosophy in general, intellectual life in Greece in the 7th and 6th centuries B.C., Thales, Anaximander, Anaximenes, Heraclitus, the Eleatics, Empedocles, Anaxagoras, Leucippus, the Pythagoreans, the Sophists, Socrates, Democritus (20 pages), Plato (50 pages), Aristotle (70 pages), the Peripatetics, Stoics, Epicureans, Sceptics, Neo-platonists, Christian Apologists, etc. 2nd German edition translated by H. E. Cushman. xv + 393pp. 5⅜ x 8.
T357 Paperbound $2.25

THE PALACE OF PLEASURE, *William Painter*
Elizabethan versions of Italian and French novels from *The Decameron,* Cinthio, Straparola, Queen Margaret of Navarre, and other continental sources — the very work that provided Shakespeare and dozens of his contemporaries with many of their plots and sub-plots and, therefore, justly considered one of the most influential books in all English literature. It is also a book that any reader will still enjoy. Total of cviii + 1,224pp.
T1691, T1692, T1693 Three volume set, paperbound $6.75

THE WONDERFUL WIZARD OF OZ, *L. F. Baum*
All the original W. W. Denslow illustrations in full color—as much a part of "The Wizard" as Tenniel's drawings are of "Alice in Wonderland." "The Wizard" is still America's best-loved fairy tale, in which, as the author expresses it, "The wonderment and joy are retained and the heartaches and nightmares left out." Now today's young readers can enjoy every word and wonderful picture of the original book. New introduction by Martin Gardner. A Baum bibliography. 23 full-page color plates. viii + 268pp. 5⅜ x 8.
T691 Paperbound $1.75

THE MARVELOUS LAND OF OZ, *L. F. Baum*
This is the equally enchanting sequel to the "Wizard," continuing the adventures of the Scarecrow and the Tin Woodman. The hero this time is a little boy named Tip, and all the delightful Oz magic is still present. This is the Oz book with the Animated Saw-Horse, the Woggle-Bug, and Jack Pumpkinhead. All the original John R. Neill illustrations, 10 in full color. 287pp. 5⅜ x 8. T692 Paperbound $1.50

ALICE'S ADVENTURES UNDER GROUND, *Lewis Carroll*
The original *Alice in Wonderland*, hand-lettered and illustrated by Carroll himself, and originally presented as a Christmas gift to a child-friend. Adults as well as children will enjoy this charming volume, reproduced faithfully in this Dover edition. While the story is essentially the same, there are slight changes, and Carroll's spritely drawings present an intriguing alternative to the famous Tenniel illustrations. One of the most popular books in Dover's catalogue. Introduction by Martin Gardner. 38 illustrations. 128pp. 5⅜ x 8½.
T1482 Paperbound $1.00

THE NURSERY "ALICE," *Lewis Carroll*
While most of us consider *Alice in Wonderland* a story for children of all ages, Carroll himself felt it was beyond younger children. He therefore provided this simplified version, illustrated with the famous Tenniel drawings enlarged and colored in delicate tints, for children aged "from Nought to Five." Dover's edition of this now rare classic is a faithful copy of the 1889 printing, including 20 illustrations by Tenniel, and front and back covers reproduced in full color. Introduction by Martin Gardner. xxiii + 67pp. 6⅛ x 9¼. T1610 Paperbound $1.75

THE STORY OF KING ARTHUR AND HIS KNIGHTS, *Howard Pyle*
A fast-paced, exciting retelling of the best known Arthurian legends for young readers by one of America's best story tellers and illustrators. The sword Excalibur, wooing of Guinevere, Merlin and his downfall, adventures of Sir Pellias and Gawaine, and others. The pen and ink illustrations are vividly imagined and wonderfully drawn. 41 illustrations. xviii + 313pp. 6⅛ x 9¼.
T1445 Paperbound $1.75

Prices subject to change without notice.

Available at your book dealer or write for free catalogue to Dept. Adsci, Dover Publications, Inc., 180 Varick St., N.Y., N.Y. 10014. Dover publishes more than 150 books each year on science, elementary and advanced mathematics, biology, music, art, literary history, social sciences and other areas.